ECO-CENTRES & COURSES

Eco-Centres & Courses

Over 150 places which offer practical courses and fun days out

Terena Plowright

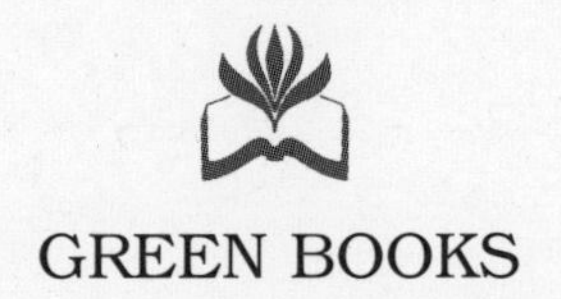

GREEN BOOKS

First published in 2007 by
Green Books Ltd, Foxhole,
Dartington, Totnes, Devon TQ9 6EB
edit@greenbooks.co.uk
www.greenbooks.co.uk

Cover photo credits:

Front cover, clockwise from top left: Aldermoor, The Ecohome,
Findhorn Foundation, Brithdir Mawr, Woodsmoke, Rippledown

Back cover, clockwise from top left: Abbotts Living Wood, Woodsmoke,
The Ecohome, Abbots LIving Wood

Cover and colour plates printed on 75% recycled paper
by Brightsea Press, Exeter

Text printed on 100% recycled paper
and bound by TJ International, Padstow, Cornwall

ISBN 978 1 903998 90 8

Contents

Acknowledgements

Although this was an exciting project, it was also rather daunting at times and so I would like to thank those who had no doubt I would complete it: Tim and Maddy Harland and John Adams from Permanent Publications (who also gave advice and support); the endless individuals that I spoke to who work in these centres; staff at The Sustainability Centre (especially Tori Melhuish); and John Elford, Jon Clift and Amanda Cuthbert from Green Books. These people allowed me to believe that I would eventually hold this book in my hands. I would also like to thank some unlikely candidates for my education in life, which has allowed me to 'see': twenty years ago 'Crass' opened my mind; my wonderful friends, who are still teaching me (particularly Heather Todd, Sarah Johnston and Carolyn Sullivan); and the many who have taught me elements of rural life (especially Gary Gamblin). Finally, I would like to thank those who gave me the rare security and comfort that allows me to be free within my soul and mind – my family, especially my mother and Bradley my brother.

This book is dedicated, with love, to those long gone:
My Dad, Oonie and Sam

Foreword

We met Terena largely through the dogs. We have an organic farm, which amongst many other things has a flock of sheep, so when Terena needed a home for one of her sheepdog's puppies, we nervously accepted the offer. Since then we have all spent many hours with the pup, Scarlet – training, watching, and, of course, getting frustrated at her attempts to get the job right. However, three years later Scarlet is fully grown, and we are the owners of a working dog, who is out on the farm daily.

It was during these times that we chatted with Terena about the effects of global warming, and we all expressed our desperation at the situation. We talked at length (with cups of tea in hand) about biomass and the difference it could make to CO_2 emissions, and also discussed other renewable sources of energy. And an exciting change has happened over the last three years: it is no longer just a few people talking about such things; it is now on the TV, on the radio, in newspapers, and has become a 'normal' topic of conversation.

People now want to know how to help the planet and how they can do their bit in the fight against rising CO_2 emissions; thankfully there are educational centres, teaching all aspects of sustainability, springing up all over the country, providing courses which help us to understand, respect and care for our planet, each other and ourselves.

These centres range from tiny family-run hobbies through to much larger organisations, and they offer a wide range of courses that really do teach people how to make a difference – however small. This book not only offers you the opportunity to find the eco-centre nearest you which runs the kind of course you want to attend, but also introduces a wide array of courses to spark your imagination, making greener living the fun and exciting experience that it should be.

Julie Walters and Grant Roffey

Introduction

What a fantastic opportunity – to spend day after day speaking to like-minded people who are spread far and wide across the country. That was the task I ended up with when I was writing this book. How humbled I felt when I learned about some of the amazing projects that people have undertaken, usually with few or no resources. The strength of the individuals when faced with true fear for their planet is inspiring, yet when there are groups of these people, the determination, imagination, willfulness and pure love for what they are trying to do becomes magical. They have one ultimate aim, yet these people are far from similar: different educations, different support networks, at home with children, on top of hills, in the middle of cities, on huge farms, working from one room, lots of money, no money, no access to computers, whole computer suites. Yet they have one overarching similarity – belief in what they are doing – and that is what makes them so good at what they do.

The flip side of this coin is not so good. For every one of those people trying to find ways to protect the planet from abuse, there are literally thousands of people ignoring what they know is happening: 'climate change', destruction of habitat, waste of resources . . . the list is huge. Yet I don't think these people deliberately turn a blind eye (or maybe I am being naïve); I think it is a lack of understanding, a lack of belief that our safe, warm comfortable world could really be taken away from us. When we look up and see a clear blue sky, it is hard to believe that there are tons of particles in the air creating a blanket around the earth; it is hard to believe there is a shortage of clean water, when we flush tons of it down the toilets every day; and it is not easy to want to believe climate change is a bad thing when the summers just keep getting better. So what a desperate situation we are left with: a few people trying to save a planet from huge changes, and many more people ignoring the situation and making it worse. We need to reverse these numbers, and this is why the centres featured in this book are so important. Most issues faced by the human population can be altered by education, and I believe that many people want to change to more sustainable lifestyles but haven't got a clue where to start.

It is very easy to sit with our heads in the sand and hope that the 'government', the 'greenies', the 'scientists' or our neighbours will sort out this crisis (and yes, it is a crisis), but actually it is not that difficult to 'sort out'. We just need to care! Once you care, you make small changes to your lifestyle, and these are what make the difference. Do YOU still leave the tap running when cleaning your teeth? Do YOU leave the TV and computers on stand-by? Do YOU still believe that biodegradable products are not as good – TRY THEM! These small changes are our most powerful defence in the fight against climate change.

Sustainability encompasses everything we do, and everything we do affects our world. Whenever we undertake any project or task, we should take into account the three main aspects of sustainable living:

- *The environment* – What surrounds us? Animals, insects, land, air, water. Are we damaging them? How do we prevent this damage?

- *Finance* – Do the numbers stack up? Is the project able to generate its own funds, or pay back loans? Is it making money by exploiting or oppressing others?

- *Community* – Are all levels of the community taken into account (the disabled, the disadvantaged, the poor)? Are we abusing anyone? Are we excluding anyone?

These three subject areas are huge, which is why it takes whole centres to try to explain their importance for a sustainable world, and the cycles within that.

This book is by no means an exhaustive list of the centres teaching sustainability across the country, but it does feature centres which are spread widely, so in each area of the country you can find a centre where you can learn aspects of sustainable living. Many centres also offer courses, which means you can meet like-minded people and learn (mostly with hands-on experience) aspects of sustainable lifestyles. Some of the centres are clearly about sustainability; for example, the Centre for Alternative Technology is an amazing place that is designed for large numbers of visitors, and teaches many aspects of sustainable living. Other places are not so obviously concerned with sustainability, yet what they teach is just as vital – for example the Country Parks. The Country Parks and Wildlife Trusts teach about our ecosystems and the tiny mini beasts that live around us, which are part of a cycle that keeps us alive; if we don't know about them and appreciate them, we will continue to destroy them with pesticides, pellets and other similar concoctions. Other places teach some simple ways to change lifestyles through projects that can be done by families. One place I would love to visit one day is Slack House Farm – what an interesting and beautiful place, what dedicated and imaginative people, and they seem to have kept their sense of humour! (See 'bits and bobs' on their website.)

We live in a beautiful world, and it is up to us to take the time to enjoy it; it is up to us to take the time to experience it; and it is up to us to treat it with care. If we don't, the world may not be destroyed, but it will change for the worse, and that change will endanger many species. Maybe the changes will make the environment too hostile for many animals to survive – and one of those animals could be the human being.

About this book

Each centre has either renewable energy technologies, or courses, or other demonstrations and examples of sustainable lifestyles. Please remember that not all the centres will have all of these. In the first part of the book I explain the basic operation of the different renewable energies and green woodwork, so you will know what you are looking at if these are exhibited at the centres you visit. The second part is a list of centres organised by region.

Below are some housekeeping points, which will allow you to use this book easily and effectively:

Finding a centre or course

To find a centre in a particular area, use the maps on pages 17-21. Each centre has been given a number, and these run sequentially through the book. To find a particular course, first use the *Index of Courses* on pages 205-214 to find a centre which offers the course you are looking for, and then find the centre's page number in the *Index of Centres* on pages 215-217.

Up-to-date information

Please always check the website of the place you are visiting. This is where you will find up-to-date information on visitor amenities and also on courses. Courses sometimes get cancelled, or new ones added.

Booking

It is also essential that you pre-book onto any courses; do not just turn up, as there may not be space for you, or equipment provided. Contact each venue in plenty of time before the course start date.

Course Dates

There are no course dates included in this directory, as these may change from year to year, or sometimes dates get altered.

Disabled access

I would like to apologise to those with a disability for not giving more information on accessibility. The centres all have different aspects to them, and each centre, area and course has different levels of accessibility. For this reason, I did not want to mislead anyone, and decided that it would be better if individuals contacted the centres to discuss access, or looked at the individual websites.

Travel information

Where possible please travel by public transport, foot or bicycle. It is not possible to access all centres in this way, but if it is, it would be good to start your visit with a green mode of transport.

Eco Centres (Encams)

Although these centres are grouped under the general term 'Eco Centre', which I think best describes them, the term should not be confused with the Encams award scheme. Some centres in this book have added their award as a point of interest; this does not mean these centres exclusively hold the award, nor does the title of the book mean all the centres included hold the award.

Other Information

It is worth checking each centre's website, as I have mostly only included information that is relevant to sustainability. Many centres have animals, trails, further courses, or other activities that may not be relevant for inclusion in this book.

Level of detail

Each centre was offered the same opportunity to provide information about its activities, and it is this information which was used to compile this book. Some centres gave more information than others, which is why the level of description differs. For example, some centres provided course titles, and some provided full course descriptions.

KEY

Recommendations for the next edition

If you would like to recommend an eco-centre for inclusion in the next edition of this book, please use the form on page 217.

RENEWABLE ENERGY

OVERVIEW

Our modern world is demanding greater and greater amounts of energy to power its 'needs'. This energy is convenient and readily available, so we have no need to think about its source. Yet each time we flick a switch, drive our cars, mow our lawns with powered mowers or even jump in the local swimming pool, we are damaging our plants, our insects, our trees, our air, our health – our entire planet.

The power we use in our day-to-day lives is derived from fossil fuels (coal, gas and oil). A fossil fuel is a fuel that has carbon locked in it (e.g. vegetation that died millions of years ago), and when we burn that fuel, it releases the carbon into the atmosphere. Carbon is dangerous because it stays in the atmosphere far above the surface of the earth – it effectively creates a blanket. As you know, a blanket stops heat escaping, and so the carbon blanket is not allowing the heat from the sun to bounce off the earth back into space, and the planet is therefore warming up.

At the moment we have two choices:

1. We can enjoy the last of the fossil fuels, see what happens, and hope the scientists, the government and the 'protesters' are wrong.

2. We can act to protect the future for children and the other species living on our planet. This means we have to 'act now', and convert to renewable energy.

Why is there now so much CO_2 in our atmosphere?

Before the Industrial Revolution everyone used renewable energy. For example, we burned logs to keep warm, or we used windmills for energy. The Industrial Revolution gave us the ability to use fossil fuels on a large scale: but the release of CO_2 into the air by industrial processes has been so great that the plants and the sea have not been able to absorb it, and it has built up in the atmosphere.

WHAT IS RENEWABLE ENERGY?

- Renewable energy is energy that will not run out, and will keep replacing itself, if we allow it to do so.

- Renewable energy does not produce carbon or release it into the atmosphere (apart from biomass, where the CO_2 released when it is burned balances that which was absorbed during the growth of the material, effectively making the process carbon-neutral.)

If you want to get renewable energy to work for you, you need to look at the potential for using the various kinds of renewable energy in your particular area. For example, there is no point in looking at power from water if you do not live near a river or the sea, and there is no point looking at wind if you are in a sheltered valley; it could be that you can use a combination of renewables to gain the best results. As fossil fuels run out, and people turn more to renewable energies, power sources will become more localised – you can even generate some electricity from your own home.

We are still learning about the potential of the various renewable energy technologies. We will hopefully continue to work towards high standards of living for everyone (not just a few), and create that wealth by using our intelligence and respect, rather than letting greed, abuse of the planet, its resources and each other dictate our actions, as we have in the recent past.

This is a great opportunity to make great changes.

UNDERSTANDING RENEWABLE ENERGY SOURCES

There are many websites that explain renewable energy in great detail, and a few are listed after each section if you want to investigate further.

Biodiesel

Biodiesel can be made from vegetable oil, animal oil or fats, and waste cooking oil (the waste oil will need treating before conversion to biodiesel, to remove impurities). It is an alternative fuel, which can sometimes be used to replace conventional diesel. Biodiesel is 'carbon neutral', which means when the oil crop grows it absorbs the same amount of CO_2 as is released when the fuel is combusted (a better balance is gained if the crop is grown organically without involving the manufacture of fertilisers). Biodiesel is rapidly biodegradable and completely non-toxic, which means that any spillages represent far less of a risk than normal diesel spillages. Biodiesel also has a higher flash point than normal diesel, and so is safer.

www.biodiesel.org.

Biomass

Biomass is biological material from living or recently living organisms. This is usually plant-based material (such as wood), but can also be both animal and vegetable derived material.

Biomass is particularly effective in the fight against CO_2 emissions because most of the power that a home uses is for heat, and when biomass is used in boilers the heat produced is CO_2 neutral. So by reducing the amount of fossil fuels used for heating homes, we are greatly reducing the discharge of CO_2. The boilers are CO_2 neutral because, although CO_2 is released into the atmosphere as the wood is being burned, the next crop of trees reabsorbs it as they grow – thus creating a cycle. It is therefore important to replant what you fell – the materials are quick to grow – or to allow coppice to regenerate.

Another major benefit of harvesting wood is the benefit to biodiversity: harvesting wood results in more woodlands being actively managed, which gives wood value and therefore encourages the care of our coppices and ancient woodland sites, as well as our commercial plantations. The owner of a small woodland can heat his own house using the timber from his own land, and with the cost of gas and oil rising, there is a considerable incentive.

A biomass boiler functions by burning organic matter. For thousands of years we have burned logs, and some farms burn straw bales or animal bedding. However, technology has now moved on, and the most popular form of biomass is wood, which can be processed to form three different fuels: logs, chip and pellet. Each one of these is unique in its use, so an understanding of the fuel is needed prior to prescribing the type of boiler.

Logs: Logs are ideal for a small building or house, either burned in the grate, wood-burning stove or boiler. There are now self-loading log boilers, but care is needed in selection of size of logs.

Woodchip: Woodchip that is used for a boiler is not the same as that used for the surface of children's playgrounds, although it is similar. The chip for a boiler is uniform in size, and the smaller the boiler the more important is the quality of the woodchip, so that the boiler feed does not become blocked, or damp wood create tar or loss of efficiency. The woodchip is delivered into a hopper and then fed into a computer-controlled boiler. Woodchip is for larger concerns, for example a block of offices, flats, or even a power station.

Wood Pellet: Pellets are sawdust compressed into pellet shapes, normally held together by the natural sap. Because of their uniform shape, pellets flow easily through pipes and are therefore suitable for domestic use. Once installed, a pellet boiler operates in a similar way to that of oil. The pellets are delivered into a container (a hopper), and then a computer regulates the amount of fuel the boiler needs to meet the required temperature for the building. A pellet boiler is suitable for a home, or for smaller office buildings.

Wood stoves: Besides boilers, you can also use biomass to fuel stoves. Stand-alone stoves are used to provide space heating for a room. These systems can use either pellets or logs (logs will not feed automatically) and some can be fitted with a back boiler. They also bring the visual effect of real wood burning into the home.

www.bioenergywm.co.uk
www.mwen.org.uk
www.swwf.info
www.wood4heat.co.uk
www.sewf.co.uk

Ground source heat pumps (geothermal): A heat pump uses a method of heating which is similar to a refrigeration unit in reverse; it transfers warmth to liquid, instead of removing the warmth from the air.

Although the temperature of the surface of the earth is constantly changing, reflecting our daily weather conditions, a few feet below ground the temperature remains much more constant because the mass of the earth can store the heat of the sun.

The system works by pumping a mixture of water and antifreeze through a pipe several feet below the surface of the ground. The heat from the earth is transferred to the water in the pipe. This heat is brought back to the building, concentrated by the heat pump, and transferred to water in a tank or used for underfloor heating.

Although heat pumps are run by electricity, they are very efficient: for every unit of electricity used to run the heat pump, about four units of heat energy are created.

They can also be used to warm water before it enters your domestic hot water heater, thereby saving on energy used. If you want to install a heat pump, you will need sufficient space outside to dig either a trench or a borehole.

www.ground-source-heating.co.uk
www.nef.org.uk/gshp

Hydropower

Hydropower is one of the oldest forms of power used by people. It is harnessed in a similar way to wind, except that a wheel is turned by either the weight of water falling onto it or by a current passing below it or through it. The energy from the moving water is thereby transferred to the moving wheel, which can then move cogs for machinery or a turbine to create electricity.

www.british-hydro.org

Solar Power

Photovoltaic (PV): Photovoltaic power comes from the space industry, where it was first used to power satellites. Now it is used to create electricity for homes and businesses.

Photovoltaic (solar PV cells) take energy from the sun to generate electricity. Photovoltaic means electricity from light, as it only needs daylight, not direct sunlight, to create power.

The amount of cells combined with the amount of sunlight hours and the strength of the sun determines the quantity of electricity produced.

www.pv-uk.org.uk

Solar water heating: Solar water heating systems are sited on the roofs of buildings to obtain the maximum heat from the sun, and this heat is then transferred into the water. They are designed to work alongside a conventional domestic hot water system (not replace it) and to take the 'chill' off the water even on cool days, which therefore reduces the amount of power needed to make the water hot. Even in the UK, a solar water heating system will supply about 60% of the domestic hot water needed in an average house over a year. In the summer nearly all the hot water can be provided this way.

www.cen.org.uk/developer/solarth.asp

Passive solar design: If a building is designed properly to use passive solar energy, it can require little or no heat from conventional heating devices. Passive solar design uses solar energy to provide the heating and lighting for a building, and sometimes it can also assist in the ventilation. There are six areas of design for consideration: orientation of the building, size of the windows and their shading, heat storage for release in the evenings, ventilation for fresh air supply, and how to insulate the building to retain the heat gathered.

Passive solar design means that the building uses less energy and so costs less to run, yet does not necessarily cost more to build.

Tidal power

Tidal power is effective because it is reliable – we are absolutely sure that the tide will come in and go out at about 12 hour intervals.

In a way similar to wind and wave, the energy is collected using turbines (although other methods are being developed). As the water or tide passes the turbines, they are forced to turn and so create power. There are many designs for different systems, but there are also concerns as to whether tidal power systems may affect estuary plant and wildlife.

www.hie.co.uk/argyll/tidal_power.html
www.environment-agency.gov.uk
www.worldenergy.org/wec-geis/publications/reports/ser/wave/wave.asp
www.esru.strath.ac.uk/EandE

Wind power

We have created power using the wind for generations – think of the old windmills that still sit on top of many hills, some of which are still in use today.

The traditional windmill uses the power of the wind to grind grain, saw wood or crush seeds, whilst a modern wind turbine converts moving wind into electricity.

The wind that travels across the UK is a huge untapped resource, and could supply a large amount of our electricity if we harnessed it. There is a wide variety of systems and sizes of turbines: for example, small ones that are used on boats to charge batteries; ones that supply a few hundred watts; and large turbines that create several megawatts. Some are so large you can climb them and stand on a viewing platform at the top.

A wind power system can stand alone with some means to store power (such as a battery), or it can be connected to the National Grid.

Several factors affect the siting of wind turbines: obviously there needs to be wind for any turbine to operate, and so the higher the turbine is, and the fewer obstructions around it, the faster the wind speed that will reach it.

There are other issues to contend with, such as planning and appropriate sizing, and yet other conditions that need to be taken into account when the siting of a larger turbine is being considered. Is it within a National Park boundary? Is it with an Area of Outstanding Natural Beauty? Is it on a flight path, or close to where aeroplanes land? And finally, will it interfere with communications and signals? None of these points will necessarily prevent the erection of a wind turbine, but they must be taken into account and permissions granted.

www.bwea.com
www.yes2wind.com

NORTH OF ENGLAND & MIDLANDS

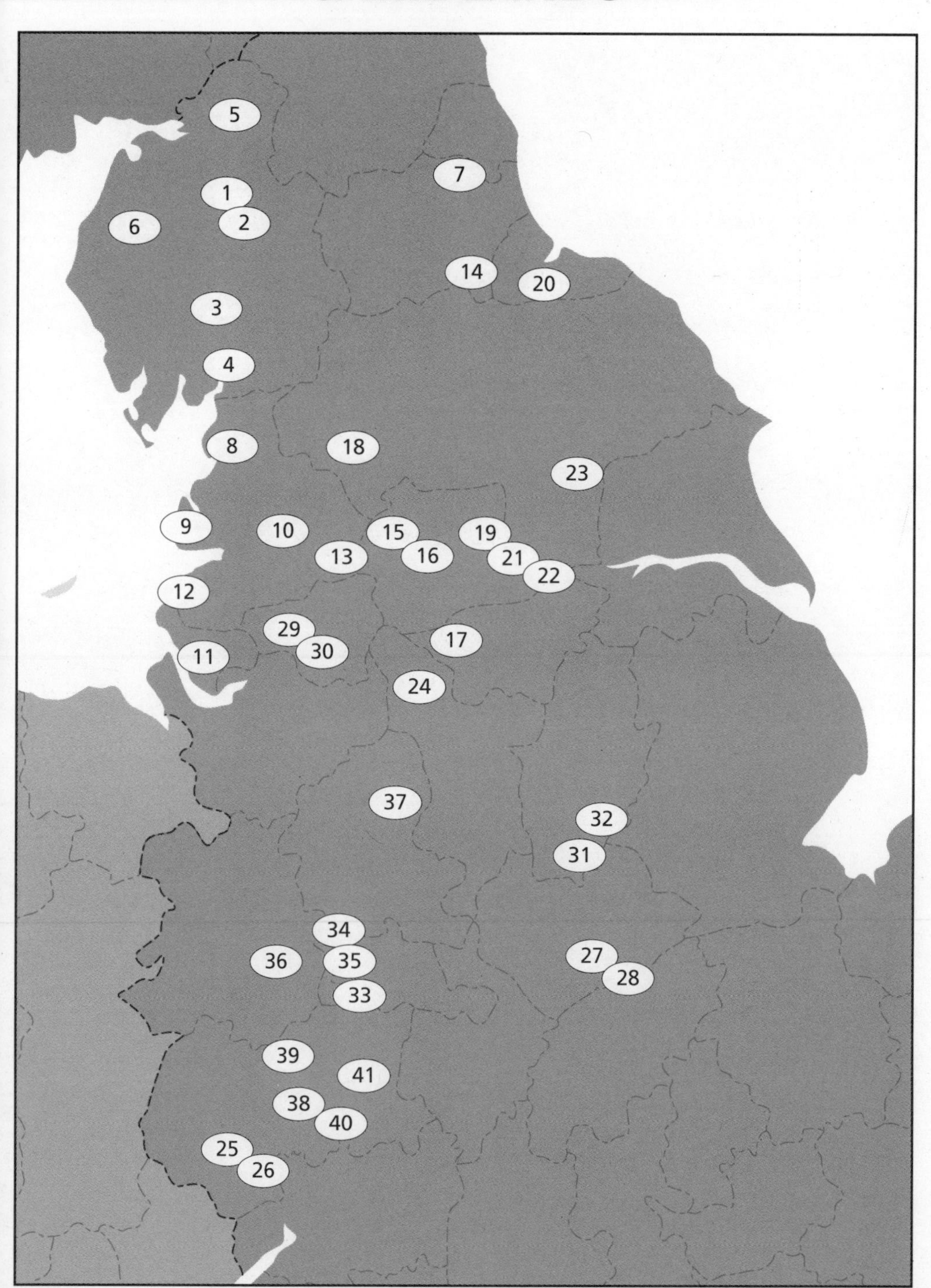

WALES & SOUTH-WEST ENGLAND

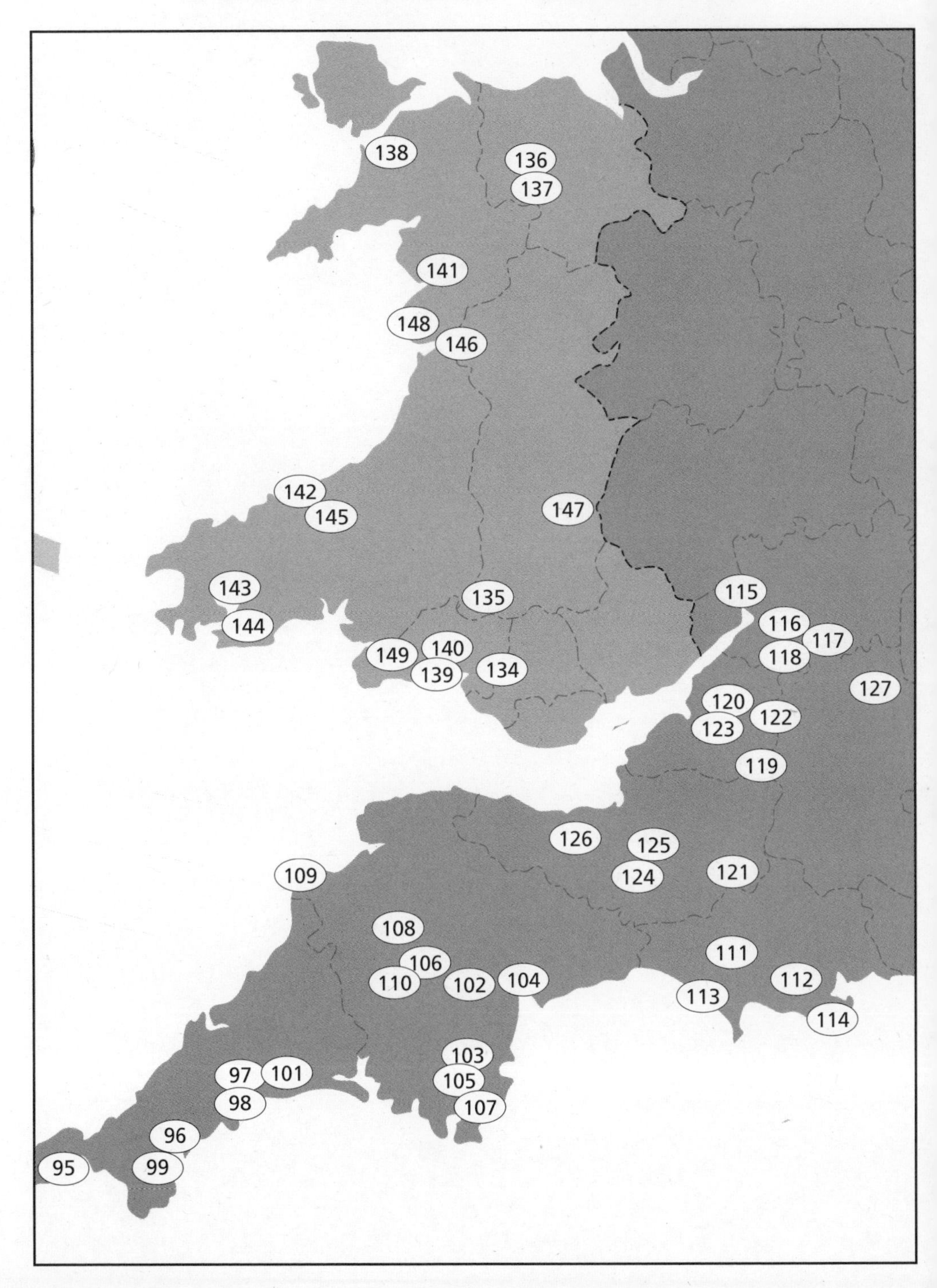

SOUTH-EAST ENGLAND & EAST ANGLIA

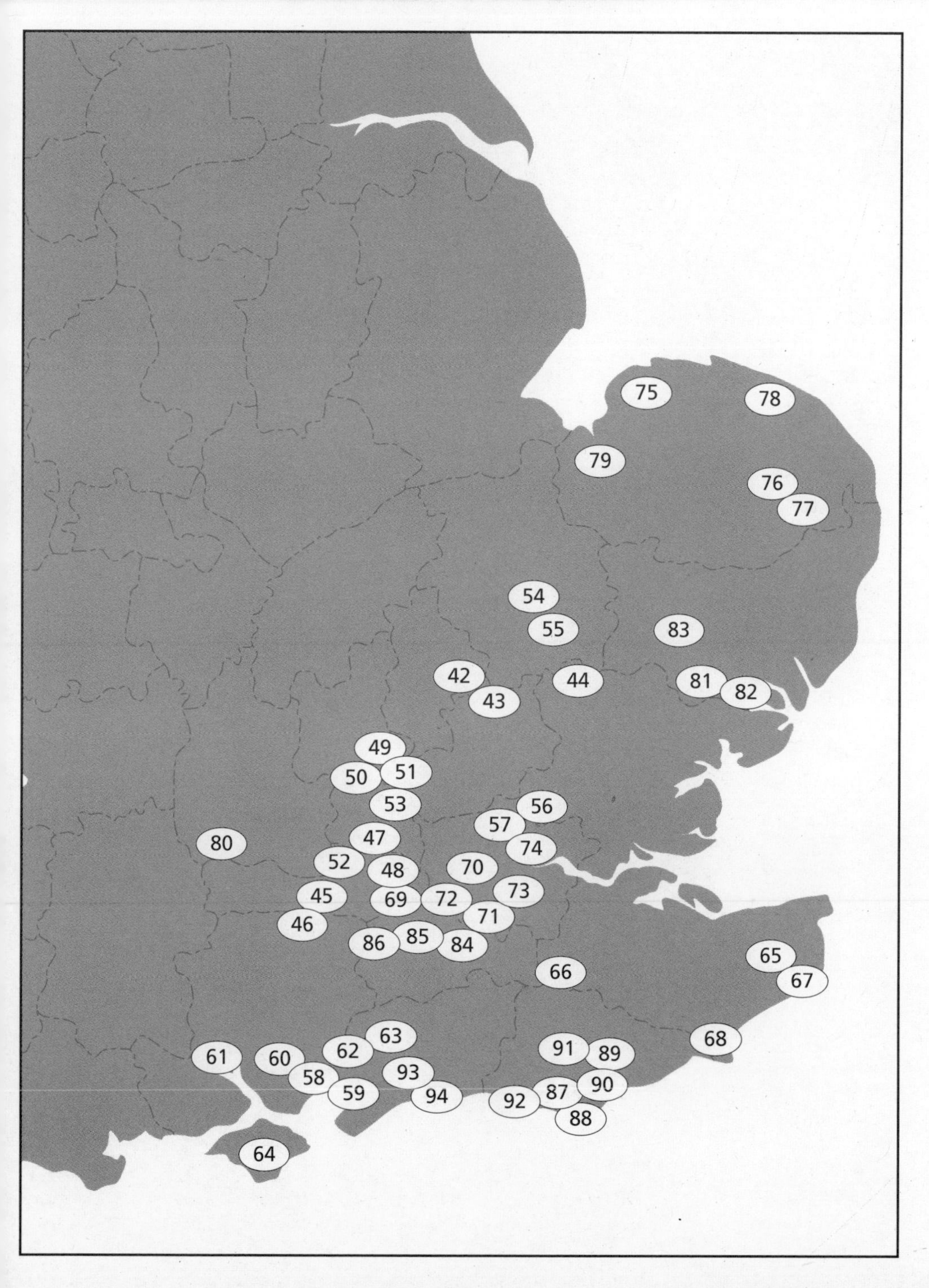

SCOTLAND

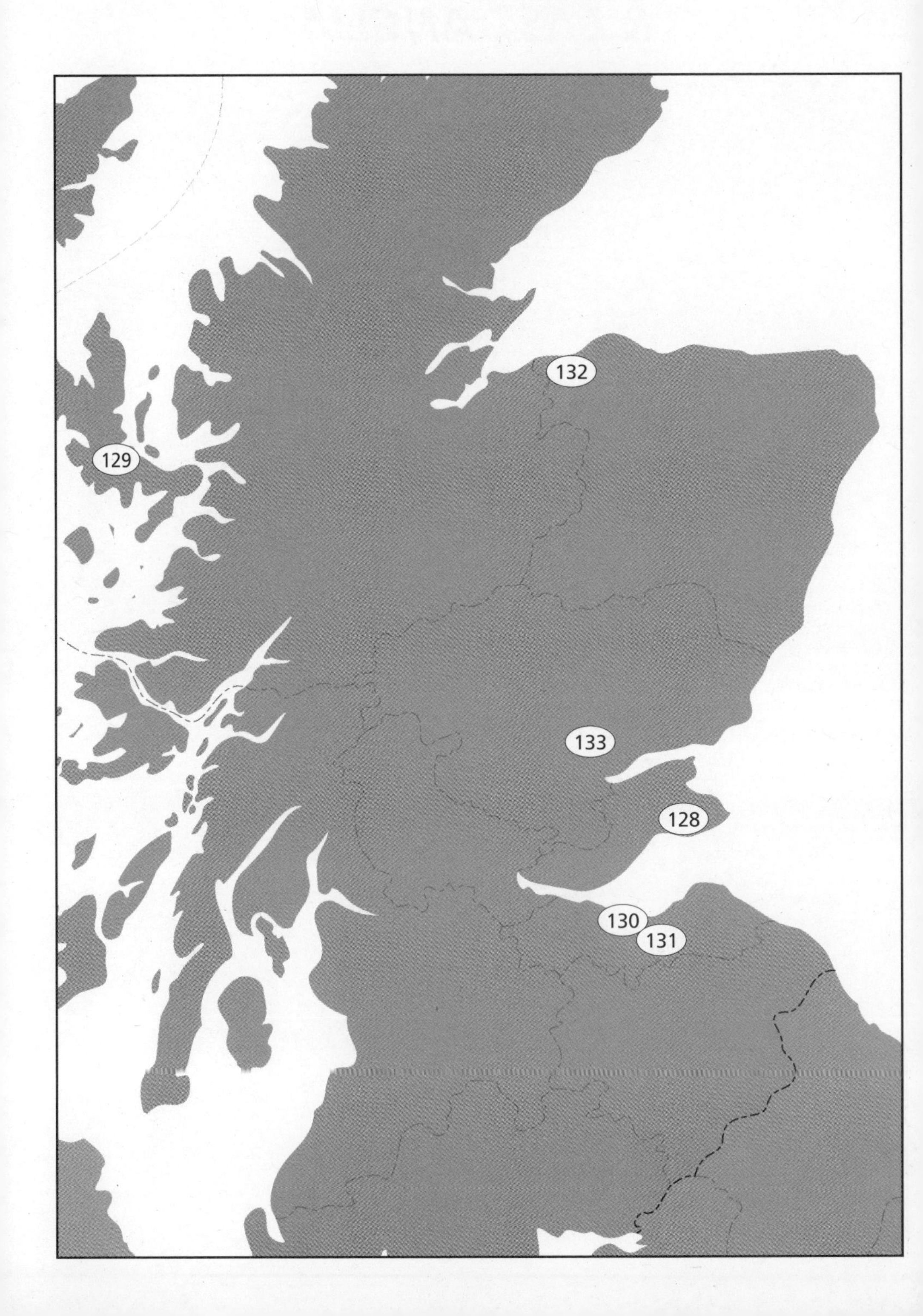

NORTHERN IRELAND & REPUBLIC OF IRELAND

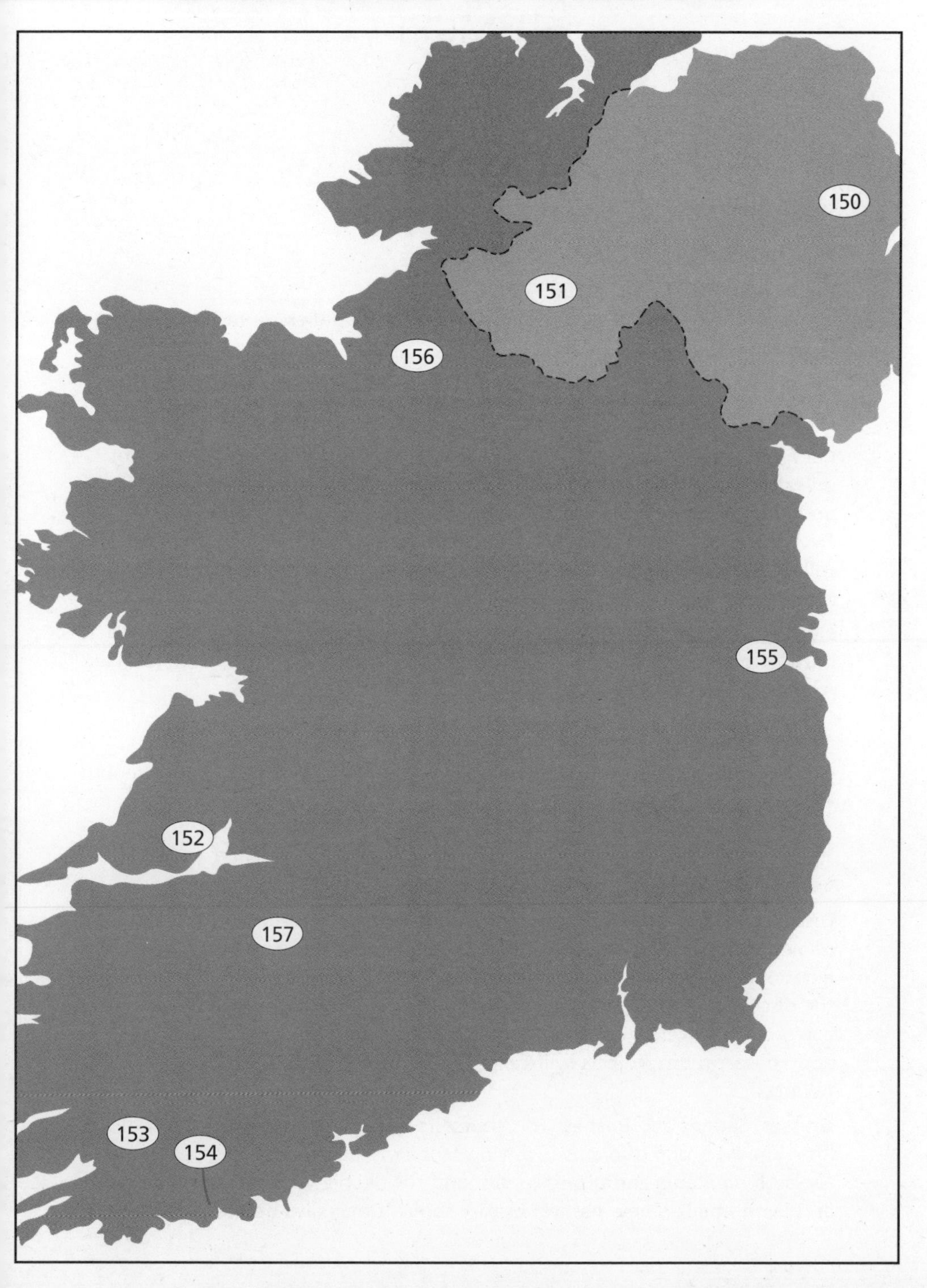

NORTH OF ENGLAND

CUMBRIA

1. Blencathra Field Centre

  CLASSROOMS, BAR, LIBRARY

Threlkeld, Keswick, Cumbria, CA12 4SG.
Tel: 017687 79601, Fax: 017687 79264.
enquiries.bl@field-studies-council.org www.field-studies-council.org

Open Not open to the general public.

Accommodation 2 holiday cottages and main accommodation (110 beds).

Directions **Train** Station Penrith 20 mins away, then catch a taxi.

Car *From Penrith:* From Junction 40 M6 follow A66 to Keswick. After about 12 miles turn off right, following signpost for Threlkeld. Go through the village and turn right onto Blease road. The Centre is located 1 mile up the hill. *From Keswick:* Follow the A66 to Penrith and M6 for 2-3 miles. Turn left and follow the signpost for Threlkeld. Go through the village and turn left onto Blease road. The Centre is located 1 mile up the hill.

Description See 'Field Studies Centres' in 'National Organisations' section at back of book. Operated in partnership with the Lake District National Park Authority. Eco Centre status.

YOU CAN SEE

Off-site field studies • Red squirrels • Living willow structures • Grounds • Weather station.

COURSES

Courses vary in length – please check website.

Discover and Explore Wild Places in the Lake District This course gives unaccompanied young people the opportunity to undertake the John Muir Award. It is a national environmental award scheme, focused on wild places. It encourages the discovery and conservation of wild places, in a spirit of fun, adventure and exploration. A variety of adventure activities, exploration of the Lake District and conservation projects will help you to meet the four challenges of the Award.

Grasses, Sedges and Rushes This course focuses on field identification of this little known group of plants in a wide range of habitats. The course is suitable for both amateurs and professionals, and for people who like to go on walks and learn about a new natural history topic. It may also be useful for people who have started this topic at a different centre and would like to see more examples of these plants in different habitats.

Family Eco Adventure A combination of adventure activities and environmental themes allow all the family to explore and enjoy the northern Lake District. Canoeing, climbing and ghyll scrambling. Mountains, lakes, sand dunes and the seashore are visited to discover as many different animals and plants as we can.

Identifying Sedges With a combination of fieldwork and supporting laboratory work, students get to grips with a variety of species. Moreover, fieldwork is at some of the best sites in the country, yielding a large proportion of British sedges. This immersion in the group through tutor-led sessions plus individual support will take the nastiness out of nomenclature and the mystery from microscopy, and leave students familiar with a wide variety of species and confident at successfully naming species in future.

Mosses and Liverworts This course provides an excellent basis on which to build further study of this group of plants, and is ideal for beginners as well as the more experienced. 70% of the species found in Europe occur in Britain, and many of those are well represented within easy reach of Blencathra. Most of each day is devoted to fieldwork which is the focus of evening talks and discussions over the microscopes.

Sustainable Uplands Management This short course is run in a workshop format and gives participants the chance to explore the issue of sustainability, taking a wide compass of policy debate. Participants seek an understanding at a local level of what sustainable uplands might look like, who would manage them and how, and what drives countryside change.

The Archaeology of the Lake District On this course we examine how over time people have changed their environment from the wildwood and fell to the present day landscapes of stone walls, parkland and whitewashed cottages. The course is open to everyone, and is suitable for anyone with an interest in archaeology, landscape history and environmental change.

Using a Flora This course is for anyone wishing to learn how to identify plants using keys in flora, for ecology professionals and for keen beginners with some knowledge of botanical terminology. You will also become familiar with the key characters of many plant families and genera.

Gardens in the Lake District

Lakeland Legends and Ghosts

People and Landscapes of the Lake District

Please also check website for further courses, e.g. art and photography, and walking.

2. Borrowdale Field Centre

 OD

Bowe Barn, Borrowdale Road, Keswick, Cumbria, CA12 5UP.
Tel: 01768 773780, Fax: 01768 774649.
borrowdale@nationaltrust.org.uk www.nationaltrust.org.uk

Open Grounds open all year. Shop and Centre open 10am-5pm, 9th March until 28th October, 7 days a week.

Directions Please call for directions.

Description See 'National Trust' in 'National Organisations' section at back of book.

YOU CAN SEE

Wood energy system • grounds.

COURSES

Various events – call for details.

3. Brockhole Field Centre

 ADVENTURE PLAYGROUND

Brockhole, Windermere, Cumbria, LA23 1LJ.
Tel: 01539 440800, Fax: 01539 445555.
education@lake-district.gov.uk www.field-studies-council.org

Open From 31st March open 10am-5pm – see website for winter times.

Accommodation Hostels and B&B nearby.

Directions **Train**: The nearest station is Windermere (2 miles).

Bus: Services 555 and 559 stop outside the gates. Contact the Travel Line on 08706 082608 for bus and train times.

Boat: Scheduled launches operate from Waterhead at Ambleside throughout the season.

Road: From the M6, exit at Junction 36 and follow the A590 and A591 to Windermere and Ambleside. The Lake District Visitor Centre is approximately half way between Windermere and Ambleside on the west side of the A591 and is well signposted.

Description Situated on the eastern shore of Windermere, the Centre is the base for the Lake District National Park's education service, which the FSC manages on behalf of the LDNPA. See 'Field Studies Centres' in 'National Organisations' section at back of book. Eco centre status.

YOU CAN SEE

Lake Windermere • Interactive exhibits • Grounds

YOU CAN DO

Guided walks • Self programmed activities.

COURSES

There is a wide range of tutored modules for primary, secondary, further and higher education students.

e.g. **Glaciation, Woodland ecosystems, Conservation and visitor management.** www.lake-district.gov.uk/understanding/education service

4. Castle Head Field Centre

   WORKROOMS, OUTDOOR CLASSROOM

Grange-over-Sands, Cumbria, LA11 6QT.
Tel: 0845 330 7364, Fax: 01539 536662.
enquiries.ch@field-studies-council.org www.field-studies-council.org

Open Not open to the general public.

Accommodation 120-bed building. Camping.

Directions 1 mile from Grange-over-Sands railway station (will collect groups with mini bus), 10 miles from M6 (Junction 36).

Description See 'Field Studies Centres' in 'National Organisations' section at back of book. Eco Centre status.

YOU CAN SEE

• Wildlife garden • Pond • Grounds • Eco centre

YOU CAN DO

• Mountain walks – learning about environment • Educational adventures – cliff climbing • Canoeing – learning about river

COURSES

Birds of Heath, Wood and Water Whether beginner or improver, this short course aims to enhance your birdwatching skills. The course features two main habitats: the Forest of Bowland for heathland and RSPB Leighton Moss for waterways, and we may also visit Walney Island for shoreline habitat. Marsh harrier, short-eared owl, avocet, waders and passerines will be our target birds. There is a chance of a dawn chorus at Leighton Moss for booming bittern and roding woodcock.

Bogs: Past, Present and Future Cumbria is an excellent location to explore peat bogs. The course includes a mix of site visits, simple fieldwork and lectures in the classroom. Most of the bogs visited are National Nature Reserves, and staff will be available to describe what is being done.The course is for those wishing to understand raised bogs, and how, for example, water needs to be managed within and around protected sites.

Butterflies and Moths The course includes countryside walks to study the special species of butterflies and moths of this beautiful area. Help with identification and information on the species and their habitat is provided. Electric observation lights are operated in the Castle Head grounds and at nearby Witherslack to attract a wide variety of moths. Morning inspection of the light traps is followed by visits to special habitats at Arnside and Meathop Moss to study butterflies. Evening talks are fully illustrated with colour slides.

Family Natural History and Adventure Climbing rocks and mountains, splashing about in lakeland streams, canoeing on a lake and camping out under the stars. Packed with fun and exciting activities. Each holiday is different.

Lakeland Spring Flowers South Lakeland is an excellent area for the study of wildflowers, and there is scope for visiting many different habitats within a fairly short distance of the Centre. The course gives an introduction to the wildflowers and includes both identifying plants and considering why they grow where they do. No previous knowledge is assumed, and we study both common and rarer varieties.

First Aid for Remote Places: Standard (HSE Certification) Standard, including HSE First Aid at Work: this progresses from the Emergency Course by introducing modules on environment, health and safety at work and illnesses. Suitable for people requiring full Health and Safety Certification under the 1981 HSE Regulations.

Further courses available e.g. **walks, first aid, wildlife** – see website.

5. Low Luckens Organic Resource Centre

Low Luckens, Roweltown, Carlisle, Cumbria, CA6 6LJ.

Tel: 01697 748186.

lowluckensorc@hotmail.com www.lowluckensfarm.co.uk

Open 7 days a week.

Entry Free.

Accommodation Self-catering.

Directions M6 motorway at Carlisle (junction 43). Nearest town is Brampton, on the A69. From Brampton go to Hethersgill, and straight across the crossroads at Hethersgill. Approximately 1 mile after Hethersgill, keep straight on (following the signpost to Shankhill and Mallsburn) where the main road bears right. After one and a half miles turn right at Shankhill school crossroads (signposted to Mallsburn and Baileyhead). After half a mile turn left at the T-junction (signposted to Roadhead), then immediately right (sign to Low Luckens). Follow the tarmac lane for one and a quarter miles, keeping left, to the farm.

Description Low Luckens Organic Resource Centre promotes sustainable farming, food and countryside, and provides opportunities for learning on the land to people of all ages and abilities. The farm is part of the Soil Association's Farm Network and produces organic beef, lamb and pork.

YOU CAN SEE

Passive solar design • Wind turbine – medium (2.5kW) • Water recycling • Rainwater collection.

COURSES

Aromatherapy Workshop • Belly Dancing Workshop • Farm Open Day • Herbal Workshop • Meditation Workshop • Mothing Evening • Music Workshop • Organic Sunday Lunch • Wildflower Walk.

6. Woodsmoke

Woodsmoke, PO Box 45, Cockermouth, CA13 9WB.

Tel/Fax: 01900 821733.

info@woodsmoke.uk.com www.woodsmoke.uk.com

Open Open Easter to Halloween for set course dates.

Accommodation Camping.

Directions Courses are conducted on a private estate in the central Lake District. Arrival details are sent on receipt of course booking.

Description Woodsmoke is a wilderness bushcraft and survival school based in the English Lake District, offering a variety of activity courses including bushcraft, wilderness survival, tracking, trailcraft, expedition skills, bow making, bush-cookery, wildfoods, canoecraft, axe work, plant identification and wilderness expeditions.

WEEK-LONG COURSES

Abo' A 6-day advanced survival course, learning to live as a British Aboriginal.

Expedition Skills Training This exciting new 7-day course is a collaborative programme with wilderness medical training and the Land Rover Experience.

Native A week-long intermediate level course of applied bushcraft skills.

Nomad A 7-day journey through the Lake District, learning the science of 'going light'.

Voyageur A 5-day open canoe expedition, experiencing the 'thrill of the paddle'.

Woodlander Six days learning the fundamental skills of wilderness bushcraft.

WEEKEND COURSES

Axe Workshop Three days learning to use the tools of the Backwoods.

Bowyer A weekend spent making a Stone Age flat bow and primitive arrows and learning how to shoot.

Lakelander A 3-day introduction to wilderness canoe craft, working in cedar-strip canoes.

Plantlore Two days discovering traditional, historical and native uses for plants, trees and fungi.

Trailbreaker A 2-day introduction to bushcraft and wilderness survival skills.

Tracker A 3-day introduction to the art of tracking and wildlife watching.

Wild food Two seasonal weekend wilderness cookery courses.

JUNIOR COURSES

13-17yrs Woodlander A 5-day course aimed at ages 13-17, based on the adult Woodlander course.

Family Trailbreaker Weekend A 2-day family introduction to Bushcraft.

OVERSEAS EXPEDITIONS

Kalahari Bushman Adventure Two-week expedition in the Namibian bush-veldt, with the Kalahari Bushman.

COUNTY DURHAM

7. Slack House Farm

Slack House Farm
Ireshopeburn
Weardale

Ireshopeburn, Weardale, County Durham,
DL13 1HL.
Tel: 01388 537292.
slackhousefarm@fleecewithaltitude.co.uk www.fleecewithaltitude.co.uk

Accommodation Bed and breakfast.

Yearly event or point of interest Various workshops throughout the year.

Directions Travelling west along the A689 in the village of Ireshopeburn, turn left on the sharp right hand bend onto Lanehill. Approximately three-quarters of a mile from the turning is a track to the left which leads to Slack House Farm.

Description Slack House Farm is an award-winning nine-acre smallholding at an altitude of 1300ft in the North Pennines Area of Outstanding Natural Beauty. The smallholding is run from 100% renewable sources of electricity, and provides bed and breakfast accommodation together with workshops on a range of rural crafts and environmental subjects. The smallholding is home to one of England's rare upland hay meadows, and has a wealth of wildlife, including black grouse.

YOU CAN SEE (BY APPOINTMENT)

Biofuel • Photovoltaics • Solar heating • Domestic wind turbine • Renewable insulation • Compost toilet • Reed bed • Recycling area • Composting.

COURSES

Please refer to website for dates or write to be included on our mailing list.

Living Willow (1 day) Learn to make living willow structures for the garden.

Natural Feltmaking (1 day) Learn the basics of feltmaking and fleece preparation using natural fleeces from Slack House Farm's rare breeds of sheep.

More Natural Feltmaking (1 day) The next step: more tips, techniques and fleeces.

Oak Swill basket making (3 days) An intensive course to introduce participants to this lost art.

Permaculture (2 days) An introduction to the theory and practice of permaculture.

Practical Green Living (1 day) A look at the practicalities of living with green technologies, including costings, grants, etc.

The Healing Garden (1 day) A look at plants in the Slack House Farm herb garden and how they can be used to provide safe and simple home remedies.

Wild Healing (1 day) Wander through the meadows and pasture of Slack House Farm and learn about the history, folklore and medicinal properties of its wealth of wild plants.

Willow basketmaking (1 day) Learn to make simple willow baskets.

FAMILY ACTIVITIES

By arrangement: family weekends to make a felt tipi.

LANCASHIRE

8. Middle Wood

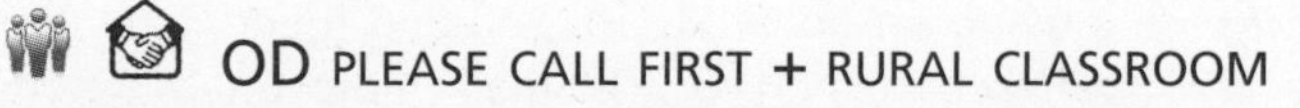

OD PLEASE CALL FIRST + RURAL CLASSROOM

Middle Wood, Roeburndale West, Lancaster, LA2 9LL.
Tel: 01524 221880.
contact@middlewood.org.uk www.middlewood.org.uk

Accommodation

The Study Centre: This ecologically designed building is powered by wind power and uses passive solar gain and solar panels to provide heat. It has a turf roof and uses sheeps wool for insulating the cavity walls. The building contains a comfortable room suitable for courses, a library area, a dining area, a kitchen, and two bedrooms each containing five bunk beds. There are two compost toilets and shower rooms. Outside there is a 'tree bog' compost toilet. The building is heated by a very efficient high mass woodburning stove.

The 10 bunk beds are provided with mattresses, pillows and blankets. Usually visitors bring their own sleeping bags or sheets. (We can provide sheets if notice is given.)

The Community Yurt: Adjacent to the study centre is the community yurt (20ft-diameter Mongolian circular tent). This is the main gathering place for the residents at Middle Wood, and is used by volunteers and visitors for accommodation.

The Camping Barn: The Roeburndale camping barn is situated in a secluded meadow on the banks of the River Roeburn, and surrounded by native woodlands. There are two floors each about 30m², and a balcony overlooking the river. The upstairs has bunk beds for 16 people: some double and some single beds.

Yearly event or point of interest Open week.

Volunteering Volunteer Weekends – maybe straw-bale building or another project.

Directions Provided when booking taken, or contact for further information.

Description Middle Wood is an environmental centre and community in North Lancashire. Ecological buildings, low impact dwellings, alternative energy systems, and woodland crafts provide the basis for human- and environmentally-friendly facilities. The Middle Wood community and Middle Wood Charitable Trust use these resources as a focus for sustainable development and, through education, help to create a future with a future. Permaculture principles and ethics act as the background for this project. The Rural Classroom space was formerly used by the Lancaster Steiner school and is located in the farmyard near the car park. The space is available for meetings and has two rooms.

YOU CAN SEE

Two wind turbines – a Proven and an old whirlwind provide power for the Study Centre and Rod and Jane's House. A 2kW photovoltaic system gives additional power to Rod and Jane's House and the farm buildings. • A hydroelectric turbine is being set up by Stuart, Ben and friends (see their dedicated website at www.fraser1.demon.co.uk). Small photovoltaic systems are planned for the School and Camping barn. These will use low energy light bulbs based on LEDs (less than 1 watt used in power consumption). • Hot water is provided from wood burning, solar panels and windmill electricity dump circuits. Where possible we use low energy appliances to reduce the need for power. The study centre has an Osier high mass woodburning stove which weighs around 1.5 tonnes and acts like a massive storage heater. This has been designed to burn very efficiently and only needs lighting for about 2 hours per day.

COURSES

Throughout the year we run a number of courses related to raising awareness of sustainability. These range from full Permaculture design courses, through youth group woodland crafts to creation spirituality.

Autumn Harvest: Herbs and Wild Food (2 days) Learn to identify, pick and process herbs and plants for medicinal and culinary purposes with an experienced herbalist.

Build your own Shave Horse for Green Wood Working (2 days) Learn to build your own shave horse to get you started with green wood working (shave horse can be taken home).

Deep Relaxation: an introduction to the principles and practice (1 day) Enjoy a day of peace and relaxation in this unique rural setting; no experience needed.

From Wool to Hat I (2 days) The weekend starts with the shearing of a sheep. You will then learn how to prepare the wool ready for knitting, spinning or weaving (washing, carding, natural dyes) to produce a final product.

From Sheep to Hat II (2 days) A continuation. Plan and design your woolly object, learn how to spin, felt or knit (skills share) to create a hat, scarf etc.

Living on the Land and the Planning System (2 days) What do you need to know about the planning system if you want to live on the land? Or build a structure, such as a compost toilet? Learn from precedence cases and share experience on how to work with the planning system.

Living Willow and Willow Weaving (1 day) Make your own living or dry garden sculpture or structure, such as an arch, trellis, obelisk, hurdle etc.

Permaculture and Sustainability Course (12 weekends over 12 months) Learn the basic design principles of Permaculture and their practical application. Permaculture encompasses not just how to grow food, but looks at how we do things and design our living and working spaces. You don't need a large garden – you can practise permaculture in your office. The course covers alternative energy sources, sewage treatment, woodland management and crafts, wildlife conservation, housing, community, alternative money design and much more.

Qigong for Health (1 day) Learn about practical ways in which to promote and maintain health and well-being or deal with health problems, working on the body's energy systems.

Reed Beds and Greywater Recycling (2 days) Learn about the principles (theory and observation of our existing systems), and apply them by constructing a reed bed and greywater recycling system.

Rustic Furniture Making I, II and III (3 x 2 days) These workshops give you the skills to design and make a piece of rustic furniture, such as a chair, stool, table, bench etc. The first weekend will focus on tools, skills and designs, followed by a weekend making your own piece under guidance.

Taiji Chi Qigong (1 day) An introduction to the principles – learn the art of nurturing our energy to maintain health and a sense of well-being in the peaceful Roeburndale Valley.

Taiji Qigong – Developing Practices (1 day) An introduction to the Chen Man-ching short form. Key postures and movements that promote balance, flexibility and control of the body, whilst promoting relaxation.

Tree Folk Law (1 day) Explore the woodland and learn about native trees.

Tree to Spoon (1 day) An introduction to green woodwork crafts in our Berber workshop.

Using Summer Herbs to make Potions, Creams, Tinctures etc (2 days) A hands-on weekend where we will identify and pick herbs and wild food, and learn how to use them for health and nutrition.

Weeds for your Needs – Spring Tonic Plants (2 days) Learn about Spring Tonic Herbs and how you can use them; wild crafting with the local herbalist Julia Russell (AMH, HMHI).

Wind Turbines – Siting, Use and Maintenance (2 days) Are you thinking about installing a wind turbine? Find out what is available and what is best for you.

9. Solaris Centre

    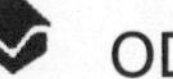 OD

Harrowside, New South Promenade, Blackpool, FY4 1RW.

Tel: 01253 478020, Fax: 01253 478021.

solaris.centre@blackpool.gov.uk www.solariscentre.org

Open Mon-Fri 8.30am-5pm, Sat and Sun 10-5pm.

Entry Free to café and exhibitions / grounds.

Yearly event or point of interest Community art / photographic exhibitions from the BBC wildlife exhibition to community art, changing every month (see website). Artificial sunshine exhibition, history of Illuminations, Sustainability exhibition, Totally Transport June 2007. Tours can be provided by appointment.

Directions From M55 follow signposts to Blackpool Airport. 100 yards past the airport, turn right at a set of traffic lights from Squires Gate Lane onto Blackpool's Promenade. Solaris Centre is located approx 300 yards further along opposite the Mirror ball. The main entrance is between the two wind turbines.

Description The Solaris Centre is an award-winning, renovated 1930s' seafront solarium that showcases sustainable design and renewable energy sources. It was developed by a group of partners from education, tourism, business, community and the local authority. In the surrounding four-acre gardens, ten artificial wildlife habitats are being created as demonstrations for schools and communities. It provides community training, tourism, employment and small business development, recycling and environmental management projects.

YOU CAN SEE (BY APPOINTMENT)

Photovoltaics • Solar heating • Passive solar design • Domestic and medium-size wind turbine • Combined heat and power • Renewable insulation • Rainwater collection.

10. The PROSPECTS Foundation

20 Cannon Street, Accrington, BB5 1NJ.
Tel: 01254 380675.
donna.neely@prospectsla21.org.uk www.prospectsla21.org.uk

Yearly event or point of interest A range of volunteer opportunities are available through practical working days and training sessions, all of which are free to residents of Hyndburn.

Directions Provided when booking or contact for information.

Description The PROSPECTS Foundation is a local community-led charity, working towards lasting environmental change in Hyndburn, East Lancashire. The role of the Foundation is to support and assist seven local community groups in developing sustainable environmental projects. The Foundation also campaigns and raises awareness of local and global issues, and has developed initiatives aimed at tackling these.

COURSES

NB: These are examples of courses that have been held previously or have been organised for the near future. As our training programme is updated on a regular basis, we cannot guarantee that these courses will be held again next year. Anyone interested in attending any of our practical working days or training sessions should contact us or visit our website for the most up-to-date events listings.

Creating a Community Wildlife Garden This course provides a basic understanding of how to create a wildlife garden, including design considerations, management considerations, plant/tree choice and care. You will also gain an understanding of the benefits of wildlife gardening to both wildlife and the community.

'Setting up an Allotment' Project A 1-day course which is ideal for those involved in developing a community allotment in their local area. The majority of this course will be theory-based work; however it will also include a visit to a local community allotment.

Tree Planting A practical, 1-day work event, which aims to improve the appearance of Hyndburn, and contribute to its environmental sustainability.

Sustainability workshops A range of half-day interactive workshops aimed at raising awareness of a variety of environmental issues such as energy efficiency, waste reduction and local biodiversity.

Willow Weaving A practical 1-day workshop which gives volunteers the opportunity to learn the age-old art of willow weaving. Participants are required to create outdoor features such as seating and fencing.

MERSEYSIDE

11. National Wildflower Centre

      OD

Court Hey Park, Liverpool, L16 3NA.

Tel: 0151 738 1913, Fax: 0151 737 1820.

info@nwc.org.uk www.nwc.org.uk

Open 7 days a week from 1st March-2nd September, 10am-5pm (last admission 4pm).

Entry Adults £3 / Children and concs £1.50 / Under 5s free. Special rates for groups. Family tickets available. Additional charges for workshops and some tours.

Yearly event or point of interest Full events programme suitable for all ages and abilities. See website.

Directions We are approx. 5 miles from Liverpool City Centre, with good rail and bus links. **Train**: Broadgreen Station is a 20-minute walk away. **Bike**: We are easily reachable from the Liverpool Loop Line cycle route, which forms part of the Trans-Pennine Way. **Car**: Follow the brown visitor signs from junction 5 on the M62.

Description The National Wildflower Centre is a charitable visitor centre, education facility, conference and function venue, promoting the creation of new wildflower habitats for people to enjoy and wildlife to flourish and develop. The visitor centre is a peaceful haven in an otherwise busy world, a family-friendly place with nature at its heart.

YOU CAN SEE

Photovoltaics • Seasonal wildflower demonstration areas.

YOU CAN DO

Volunteering • Children's play area • 160 metre long rooftop walkway.

12. Southport Eco Visitor Centre

    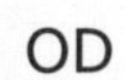 OD

Esplanade Park and Ride, Southport PR8 2BZ.

Tel: 0151 934 2713, www.southportecocentre.com

Open 7 days a week 8am-6pm.

Entry Free (£1 parking). Access to everything except the classroom. Courses need to be pre-booked.

Directions Head for Southport. From the North A565 at Crossens join the Coastal Road and follow signs for Esplanade Park and Ride. From the South A565 to Woodvale Aerodrome, then turn left onto Coastal Road, then follow signs for Esplanade Park and Ride. From the East A570 into Southport, follow signs for Esplanade Park and Ride. Look for the wind turbine.

Description The Southport Eco Visitor Centre is a unique gateway for travel into Southport, offering a practical demonstration of sustainability and inspiring visitors to consider our impact on the environment; it offers free access to information about the environment with a particular focus on climate change, energy, water, travel and tourism.

YOU CAN SEE (BY APPOINTMENT)

Thermal mass • Specialist ventilation • Exchange heat pump • Solar heating • Passive solar design • Wind turbine – medium-sized • Renewable insulation • Rainwater collection.

YORKSHIRE

13. Amazon Nails – Strawbale Futures

amazonails®

Hollinroyd Farm, Todmorden, OL14 8RJ.
Tel: 0845 458 2173, Fax: 0845 458 2173.
info@strawbalefutures.org.uk www.strawbalefutures.org.uk

Open Office-based, working out in the community. Please pre-book courses and school visits (we will come to you).

Yearly event or point of interest Courses offered throughout the year on various aspects of straw-bale and natural building on live building sites.

Description Amazon Nails – Strawbale Futures is a not-for-profit social enterprise operating for the benefit of the community. It offers training, design and consultancy in straw-bale building, low impact foundations and natural building materials and techniques including lime renders and clay plasters.

COURSES

Courses are generally attached to live building sites, or can be tailor-made. For example:

How to construct a small building in less than a week – Hands-on practical work and theory presentations including history; planning and building control issues; global and environmental context; methodology; design; characteristics of straw; costing and maintenance. Practical and theory-based courses on lime renders and clay plasters, as well as low impact foundation choices and materials and sustainable roofing options. We can also give fully illustrated talks and presentations, and offer bespoke workshops and training days for groups, or team-building events.

14. Clow Beck Eco Centre

Clow Beck Eco Centre, Old Spa Farm, Croft-on-Tees, Darlington, DL2 2TQ.
Tel: 01325 721778.
info@clowbeckecocentre.co.uk www.clowbeckecocentre.co.uk

Open Bookings only.

Directions See website for full directions.

Description Clow Beck is an example of working with nature's core elements of earth, water, air and sunlight. A centre dedicated to teaching people about the natural life cycle of food. A place to learn how to respect, save and recycle and inspire for everyone to work together to make a difference to our world. It has a unique straw-bale building on a beautiful farm site and offers a range of hands-on courses and events aimed at demonstrating an alternative and sustainable way of life.

COURSES

Straw-Bale Building You will learn basic methods of building with straw, and get an overview of infill and load bearing types of construction.

Limecrete Course This course shows how to lay a sustainable floor in a timber-frame building. You will learn about using lime for floors and walls, and how to create an insulated floor suitable for underfloor heating.

Other courses: Gardening, Food miles, Cooking, Composting, Woodworking and many more; see website.

15. Gibson Mill

  + EDUCATION ROOM

Hardcastle Crags, Estate Office, Hollin Hall, Crimsworth Dean,
Hebden Bridge, West Yorkshire HX7 7AP.
Tel: 01422 844518, Fax: 01422 841026.
hardcastlecrags@nationaltrust.org.uk
www.nationaltrust.org.uk

Open *Grounds* All year, 7 days pw.
Gibson Mill 1st March-30th Sept 11am-4.30pm Sat and Sun
1st Oct-28th Feb 11am-3.30pm Sat and Sun
Muddy Boots Café 1st March-18th April Sat and Sun
18th April-30th Sept Tues / Weds / Thurs, Sat and Sun
1st Oct-28th Feb 11am-3.30 Sat and Sun

Directions Please call for information. Visitors are encouraged to come on foot, by cycle or public transport. Car parking and cycle racks available at Clough Hole car park on Widdop Road and Midgehole car park on Midgehole Road. During busy times, limited car parking leads to heavy congestion.

Description Gibson Mill embodies the Trust's sustainability message: this 200 year old mill is 100% self-sufficient in heat and electricity, as well as water and waste treatment. See 'National Trust' in 'National Organisations' at the back of the book.

YOU CAN SEE

Water mill (generating all own power) • Recycling all own waste • Photovoltaics • Solar hot water • Biomass harvesting • Biomass boiler • Ceramic stove • Local water source • Water efficiency • Building constructed in local stone • Recycled materials and timber • Design to maximise light in café • Insulation • Compost toilet • Human-powered lift.

COURSES

Please send s.a.e. for details of events.

16. Hebden Bridge Alternative Technology Centre

       OD

Hebble End Mill, Hebden Bridge, HX7 6HJ.
Tel: 01422 842121, Fax 01422 843141.
info@alternativetechnology.org.uk www.alternativetechnology.org.uk

Open 10am-5pm (12-5 Sat, 12-4 Sun and bank holidays)

Entry Free. Please pre-book guided tours.

Accommodation Nearby (Woodcraft folk hostel available for hire).

Yearly event or point of interest Big Green Week – where we showcase what people here are doing and the best of what is happening elsewhere.

Directions On the canal towpath in the centre of Hebden Bridge, behind the Co-op supermarket.

Description The ATC works from a strong base within the local community, provides inspiration, accessible information and advice, practical, innovative and exciting examples of sustainability, enabling people to improve all aspects of their lives and their environment.

YOU CAN SEE

Biomass boiler • Exchange heat pump • Photovoltaics • Solar heating • Renewable insulation • Rainwater collection • Sustainable retrofit of some units using recycled materials, eco paint, sheep's wool insulation, recycled carpet tiles etc.

COURSES

Building a solar water panel or wind turbine.

Drystone walling and other countryside skills.

Hedge-laying Join a working group and learn about hedge-laying – whilst improving the hedges around the Calder valley.

Mosaics from reclaimed and found objects.

Papercrafts, felt making and other eco crafts.

Plastics recycling workshops For artists/craftspeople or for people wanting to set up their own community factory.

Renewable energy basics Wind, hydro, solar water and PV, biomass.

Weaving with willow and other hedgerow materials

Wildlife workshops Minibeasts, frogs and toads, wildlife surveys.

Wind workshop Make a wind turbine demo that produces electricity.

Check our website for details as the list of courses changes and grows.

WORKSHOPS FOR SCHOOLS

Plastic recycling.

Crazy shapes and keyrings Collect, sort and process recyclable plastics and make your own unique key fob, fridge magnet, badge or drinks coaster. Curriculum links: science, art, design and technology.

Notebooks galore Make your own notebook from recycled plastic. Curriculum links: science, art, design and technology.

Eco footprints Explore the consequences of lifestyle choices on the environment and look at ways of reducing students' (and schools') 'footprints'. Curriculum links: science, geography, citizenship.

Food miles Play the food miles game to discover just how far your food has travelled and design a low-impact menu. Curriculum links: science, geography, citizenship.

Paper recycling Pulp, process and colour waste paper and turn it into useable sheets of paper. Curriculum links: science, art, design and technology.

Renewable energy Investigate energy use, consider fossil fuels versus renewable energy, and look at the practical application of renewable energy models, including solar PV, wind, water and biomass. Curriculum links: science, technology, geography, maths.

Windpower Understand the basics of wind power and make working wind power models to generate energy. Curriculum links: science, geography, technology.

17. Heeley City Farm

 OD

+ PLAY AREA

Richards Road, Sheffield, S2 3DT. Tel: 0114 2580482, Fax: 0114 2551400.
Energy@heeleyfarm.org.uk www.heeleyfarm.org.uk

Open 7 days a week 9am-4.30pm.

Entry Free (donations welcome). Please pre-book – guided tours of the Farm and Energy Centre are by appointment only, and incur a charge.

Yearly event or point of interest Heeley Festival, held on the nearest Saturday to 21st June, hosts over 7,000 visitors each year.

Directions By road: Take A61 south from City Centre. After approx 1.5 miles turn left at Halfords, and second right.

Train/bus: From Sheffield station, take the 53 bus, get off just after Halfords, turn back the way you came, and turn right at the lights. Take the first right and the first left, and a 150-yard walk up the road (past our Energy Centre) will take you to the café.

Description Heeley City Farm is a community-based, environmental visitor and training centre based on a 5-acre urban site among Victorian housing. The Farm is home to a number of rare-breed animals, and boasts a peat-free garden centre, a vegetarian café and a Sustainable Energy Centre. It celebrated its 25th anniversary in 2006.

YOU CAN SEE

Thermal mass • Exchange heat pump • Photovoltaics • Solar heating • Passive solar design • Wind turbine – medium • Solar wood-drying kiln • Trombe thermal wall • Straw-bale building • Timber-framed building • Renewable insulation • Rainwater collection • Sustainable energy demonstration centre • Friendly animals.

COURSES

Amenity Horticulture (leading to NVQ levels 1 and 2 in Horticulture) Covering a wide range of skills including the use of machinery, plant knowledge and nursery practice, this course is ideal for those contemplating a career in Horticulture. Part time: 1-2 years.

An introduction to solar water-heating Presented with David Garlovsky of the Schools and Homes Energy Education Project, this course covers the basics of solar water heating through a mix of theory and practical. During the course each student will construct a small working model solar water heater. Past students have gone on to use these for camping, on the allotment, or simply to surprise their friends with the power of the sun. No previous experience necessary.

An introduction to straw-bale building Learn the theory behind straw-bale building, and learn the practicalities by building a corner of a load-bearing straw-bale building. The course includes the opportunity to mix and apply lime render.

An introduction to timber-frame building Another mix of theory and practical sessions. We look at the background to the Walter Segal method of timber-framed building, and will construct and erect simple frames for a single-storey building.

Environmental Conservation Part time (48 weeks). Provides students with the range of skills necessary for involvement in voluntary or professional conservation work.

Green Refurbishment and alternative energy A seven-week evening course, 6.00-9.00pm Wednesdays. A theoretical course covering energy-efficient refurbishment and limiting environmental impact. Includes a look at renewable energy options for household use. The course will be held in the eco-refurbished South Yorkshire Energy Centre, and includes discussions of the improvements carried out to that building, and a visit to another building of environmental interest.

Organic Food Growing (leading to Open College Network Level 2 in Horticulture) Part-time: 1-2 years. Provides students with the skills and knowledge to grow successfully a wide range of food crops in an organic and sustainable manner.

Seminars 4-6pm on one Friday each month. A series of short sessions on topics around renewable energy, sustainable building, energy efficiency and affordable warmth. Contact 0114 2584574 for details.

Coming soon: **An Introduction to Photovoltaics, Sustainability for Beginners, Working with Lime.**

Bespoke courses on a wide range of sustainable building topics run on request. Potential clients include local authorities, housing associations, architects, builders, quantity surveyors, community groups and many more.

18. Malham Tarn Field Centre

  5 WORKROOMS

Settle, North Yorkshire, BD24 9PU
Tel: 01729 830331 Fax: 01729 830658
enquiries.mt@field-studies-council.org www.field-studies-council.org

Open Not open to the general public.

Accommodation 70 beds in building.

Yearly event or point of interest The Malham Tarn Research Seminar. A series of talks, displays and discussions about landscape, conservation and the future, hosted and supported by the Field Studies Council at Malham Tarn Field Centre.

Directions

Train: 7 miles from Settle (get a taxi, or groups can be met by the minibus).

Car: From Malham (Skipton, Leeds, M1, A1), bear left at the Buck Inn. Drive up a long winding hill and along the flat for 3 miles to a crossroads. Go straight ahead here, past High Trenhouse (Centre for Management Creativity) and down to a junction. Turn right and drive with the wood to your right for ½ mile. Ignore the gated track and watch out for a sharp right hand: turn down a slope at the end of the woodland.

NB: Do not take the road from Malham that goes past the Listers Arms Hotel, as this becomes a gated track.

Description See 'Field Studies Centres' in 'National Organisations' section at back of book.

YOU CAN SEE

Weather station • Information boards • Grounds

COURSES

Arboriculture and Bats: a Guide for Practitioners This course, developed by the Bat Conservation Trust, Arboricultural Association and Lantra Awards, is aimed at arborists, and will give sufficient knowledge to allow them to carry out tree works taking into account the potential effects on bats and their habitats. Course content includes: basic bat biology and ecology; bats and the law; potential tree roost recognition; where to go for help; emergency procedures and practical exercises.

Bird Survey Techniques Birds are often seen as a good measure of the health of the environment, and the data produced from accurate, long term surveys are essential to measure population trends. This is a course for the keen amateur ornithologist or wildlife professional who would like to learn some basic bird survey techniques.

Bird Watching in the Dales A short course, aimed at both beginners and those who wish to extend their knowledge of birds. Birds studied include common garden and woodland species, and also some of the birds that are special to the uplands of the Yorkshire Dales. The course covers: basic identification skills by sight and by song; discussion on how birds have adapted to their habitat; exploration of different habitats including deciduous woodland, upland rivers and freshwater lakes, limestone grassland, peat bog and managed grouse moor.

Fantastic Felt This course covers basic technique and provides opportunities to experiment with the incorporation of fragments of fabric and threads and other decorations. Students can make a small one-piece bag using a basic resist technique and more experienced students may concentrate on using resists if they wish.

Grasses, Sedges and Rushes The Yorkshire Dales is an area of exceptional interest with heather moorland, limestone grassland, ash and oak woodland, high altitude fens and wetlands. There are even arctic-alpine species which used to thrive when conditions were much colder, and now only survive in a few sites. The course includes an introduction to the more common grasses, sedges and rushes; an introduction to flowering parts and vegetative characters helpful in identification of these plants; and examination of these under a microscope.

Mammal Identification This course teaches you how to identify all of Britain's land mammals, from sightings and field signs (including calls, droppings, feeding remains and footprints). Led by enthusiastic and knowledgeable tutors, the course offers lectures, slide shows and evening activities, as well as hands-on experience doing owl pellet analysis and Longworth live-trapping for small mammals.

Rag Rug Making Some of you may remember Granny making rag rugs years ago from old sacks and the family's discarded clothes. This important domestic skill was almost lost in the years following the Second World War, but since the early 1980s a number of important artist-makers have taken the traditional techniques and brought them up to date. With today's emphasis on recycling, contemporary ragwork with its vibrant colours and unusual textures is well worth rediscovering.

Slugs and Snails Slugs and snails are a very obvious component of many ecosystems, yet they are often ignored by naturalists who carry out biological surveys. The course shows you how to find, identify and interpret this exciting and accessible group of animals.

Spinning for Beginners and Improvers Learn how to choose, sort and prepare a fleece to make yarn suitable for knitting and take away a hank of yarn – enough to knit a bobble hat or gloves. Spinning will start on a drop spindle and progress to a wheel. Care of your yarn and the wheel will also be discussed. Experienced students will be able to spin different fleeces and also tops, and learn new techniques.

Spiders: Ecology and Identification This short course is designed for general naturalists, teachers, countryside managers, wardens and rangers interested in learning more about this important and often maligned group of invertebrates. Fieldwork sessions introduce a range of techniques for finding and collecting spiders. Laboratory workshops enable both novices and experienced participants to practise their identification skills. In the evenings, brief lectures and illustrated talks will provide an overview of spider ecology and conservation issues.

Wild Flowers for Beginners Malham Tarn is a fascinating area with a wide variety of different habitats – woodlands, wetlands, grasslands, limestone pavements – giving a wonderful diversity of wildflowers within walking distance of the Centre. This short course is an opportunity to look at this diversity with an experienced botanist and ecologist, noting what grows where and what grows with it.

Further courses on website, e.g. **Limestone Flora.**

Meanwood Valley Urban Farm

  OD E2E LEARNING ME / ADULTS WITH DISABILITIES

ad, Leeds

29759, Fax: 0113 2392551. education@mvuf.org.uk

Open 7 days a week 10am-4pm. Café closed Mondays.

Entry £1 adults/50p children over 12 (under 12s free). Please pre-book school projects, tours, and the hire of the EpiCentre.

Yearly event or point of interest Spring: Plant Fayre/Healing Fayre. December: Santa visits the farm + evening carol concert.

Directions **Train**: Nearest railway station Leeds City – farm is three miles from station. **Bus**: Public transport from city, bus numbers 51/52. **Bike**: For local cycle routes visit www.sustrans.co.uk.

Description A City Farm established in 1980, with a purpose-built environment centre (EpiCentre), recognised by CABE as one of the best examples of park architecture. Featuring turf roof, passive solar heating, composting toilets and reed bed drainage system. Located in the inner city, spanning a 14-acre site with gardens, nature area, café and shop. Livestock include sheep, donkeys, cows, pigs, goats, poultry. The Farm provides services to the community including environmental education projects to schools, holiday playschemes, training and development programmes to unemployed and disabled, allotment to local community.

YOU CAN SEE

Passive solar design • Photovoltaics • Turf roof • Timber-framed building made from locally sourced timber (larch) • Warmcell insulation • Compost toilet • Rainwater collection • Greywater system • Reed bed • Paper recycling.

20. Nature's World

Nature's World

      OD PLAY AREAS, INTERACTIVES

Nature's World, Ladgate Lane, Acklam, Middlesborough, Tees Valley, TS5 7YN. Tel: 01642 594895, Fax: 01642 591224.

mail@naturesworld.org.uk www.naturesworld.org.uk

Open 7 days: summer 10am-5pm, winter 10.30am-3.30pm.

Entry Varies. Please pre-book tours.

Accommodation Nearby.

Yearly event or point of interest Monthly farmers' market and craft market, on last Sunday of every month, 10am-2pm.

Directions Follow the 'Brown signs' from A19 onto A174 (south of Middlesbrough).

Description The North of England pioneering eco-experience, featuring 'Future World' with geothermal, solar and wind power. Also unique Eco trail, River Tees trail, Organic Garden trail and Wildflower trail. Walk-through compost heap interactive! Tearooms, shop and play areas.

YOU CAN SEE

Biomass production – woodchip (demo) • Thermal mass • Exchange heat pump • Photovoltaics • Passive solar design • Wind turbine – domestic • Wind turbine – medium • Geothermal – heat and cool • Earthship • Straw-bale building • Modern earth building • Timber-framed building • Renewable insulation • Rainwater collection • Greywater system • Reed bed • Green roof • Hydroponicum.

YOU CAN DO

Volunteering.

COURSES

Art (watercolours) • Beekeeping • Falconry • Horticulture – City and Guilds Skills Tests • Pottery • Wildflowers.

21. Skelton Grange Environment Centre

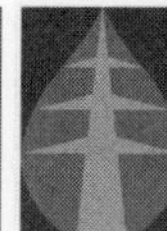

Skelton Grange Road, Leeds, LS10 1RS. Tel: 0113 243 0815, Fax: 0113 235 0046.
skelton@btcv.org.uk www.skeltongrange.org.uk

Open Open to the general public ONLY by pre-booking. Open Mon-Fri 9am-5pm.

Entry Varies.

Yearly event or point of interest Please check website for range of voluntary opportunities, community and school activities.

Directions 2 miles south of Leeds City Centre following A639, turn left up Skelton Grange Road, cross bridge at top and turn left onto Knowsthorpe Lane. Centre is on right. See website for more detailed directions, cycle route and up-to-date bus numbers.

Description A partnership between BTCV and National Grid, with support from Leeds City Council, Skelton Grange Environment Centre is based in an award-winning environmental building set in a 5-acre wildlife area. It has a classroom and meeting facilities, and a safe, fully accessible wildlife area including ponds, young woodland, spring/summer meadows, adventure trail, willow tunnel and dome, an organic vegetable garden, a riverside path and a large mature hedge. Their role is to promote environmental awareness in the community through working with both children and adults from schools, playschemes, businesses, voluntary, government and health organisations.

YOU CAN SEE (BY APPOINTMENT)

Exchange heat pump • Photovoltaics • Solar heating • Passive solar design • Wind turbine – medium • Timber-framed building • Renewable insulation • Rainwater collection • Greywater collection.

COURSES

The Project provides advice and training for teachers and conservation workers, and runs several training courses as part of the BTCV Training Programme and for the Environmental Trainers Network (ETN).

YOU CAN DO

Volunteering.

CHILDREN'S ACTIVITIES

Summer playscheme.

22. Womersley's

Womersley's Ltd. Walkley Lane, Heckmondwike, WF16 0PG.
Tel: 01924 400651, Fax: 01924 403489.
info@womersleys.co.uk www.womersleys.co.uk

Open Not open to the general public.

Description Since it was established, Womersley's has been providing a wide range of building materials for the refurbishment of historic buildings. The natural materials supplied for building, plastering, rendering, insulating and decorating your home allow the exchange of water vapour and control the relative humidity of our rooms and benefit our health. The products sold are for historic building refurbishments and eco-friendly new builds. There is also a range of courses designed for architects, builders, developers, structural engineers, surveyors, historic homeowners and DIY enthusiasts. These courses are for those people who want to learn more about traditional historic building practices and use their eco-friendly building materials.

COURSES

Decorative Paint Techniques using only Natural Paints The 1 day course only uses paints and materials that don't damage human health, and is ideal for those people who have become over-sensitive to modern synthetic petrochemical-based paints. The training shows you how to create decorative effects using only natural pigments, paints and waxes; practically based and suitable for beginners and those with some experience, it covers: Lasur glazing technique, ragging, bagging, dragging and sponging; semi-opaque casein colour washes, wood stains and colour washes for furniture and woodwork, textured wall finishes and stucco/marbleised finishes.

How to Build a Wind Turbine This week is run by Hugh Piggott and hosted by Womersley's and Amazon Nails. A full week in the workshop constructing wind turbines from scratch.

Plastic stone repair and Grouts for Historic Building Restoration Plastic stone repair promoted by the Society for the Protection of Ancient Buildings is a recognised method of repairing historic buildings whilst retaining the maximum amount of the historic fabric. The day will also look at the latest progress in the development of grouts, used to retain historic walls and structures, and provide you with an update on materials and techniques for both these basic restoration methods. The course covers: the arguments for and against replacing historic stone and brickwork; choosing the right material for plastic stone repair and good working practice; grout mixes and their compatibility with old buildings; practical demonstrations and workshops.

Renders, plasters and surface finishes This 1-day course provides advice to those who wish to restore their period property and demonstrates how breathable renders and plasters can be efficiently applied to new buildings. It includes: an introduction to breathable structures, lime and clay plasters and lime renders, historic and environmentally sound paint finishes, practical demonstrations and workshops.

Selecting The Right Lime Mortar A practical day devoted not only to the history of lime but to the specification and choice of lime mortars for conservation, restoration and new build work. The day covers: the technology of lime – fat limes, hydraulic limes and cements; mortar sampling and replacement mortars; choosing the right aggregates for lime mortars, and practical demonstrations and workshops.

Straw-Bale Building A 1-day introduction through practice and theory. Most of the day is spent erecting a straw-bale building; please dress appropriately. The course covers: different methods and techniques; planning and Building Regulation issues; how to do it and applications in the UK climate. The training is provided by Barbara Jones of Amazon Nails, a well-known international straw-bale builder.

Sustainable and Eco-Friendly Building A day looking at the latest progress in the development of using natural and more sustainable building materials in the construction of new homes. The course covers: why we should build more sustainable buildings; healthy buildings = healthy people; lightweight concrete; wall, roof and insulated floor construction; non-toxic finishes and high performance glazing; practical demonstrations and workshops.

Understanding Natural and Mineral-Based Paints A day of theory and practice devoted to understanding how paints work. It covers: an introduction to traditional and modern mineral-based paints; the plant/paint cycle; film-forming paints and vapour permeability; demonstrations showing the use of lime washes; distempers and glass-based and natural resin wall paints.

23. York Environment Centre

      OD MON-FRI AND SAT

York Environment Centre, Bull Lane, York YO10 3EN.

Tel: 01904 411821, Fax: 01904 411821.

info@stnicksfields.org.uk www.stnicksfields.org.uk

Open Open to the general public Mon-Fri 9am-5pm, Sat (usually 10am-2pm but please ring to confirm) and for events.

Entry Free. Please pre-book guided tours and room hire.

Yearly event or point of interest Big Recycle and Junk Swap Day, first Saturday in June. Apple Day, first Saturday in October.

Directions **Walk/bike**: Situated off A1079 just one mile's walk or cycle ride from the centre of York (Sustrans cycle route 66). **Bus:** Regular buses from the railway station.

Description The York Environment Centre is run by Friends of St Nicholas Fields who have transformed a former landfill site into a Local Nature Reserve and the Environment Centre. From this base, the environmental charity provides inspiration, advice, practical examples and services to enable people in York and beyond to move towards a sustainable future. They run a zero-emission kerbside recycling service and offer free composting advice. They sell a growing range of composting and recycling equipment and some books.

YOU CAN SEE (BY APPOINTMENT)

Thermal mass • Photovoltaics • Solar heating • Clearview wood-burning stove • Passive solar design • Wind turbine – medium (2.5kW) • Excellent insulation • Compost toilet • Rainwater collection • Sedum roof.

YOU CAN DO

Volunteering.

CHILDREN'S ACTIVITIES

Running Wild Club (run by Yorkshire Wildlife Trust) every Wednesday 4-5:30pm, fun outdoor activities for children aged 8-13.

 **Adult courses**  **Café / Restaurant**  **Composting**

 **Garden**  **Grounds open to the public**  **Meeting facilities**

 **Recycling**  **School visits**  **Shop** **Trails**

 **Visitor Centre** **OD Open Daily (or on days specified)**

MIDLANDS

DERBYSHIRE

24. Woolgathering

  OD WEDS-SUN

Eyam Hall Craft Centre, Main Rd, Eyam, Hope Valley, Derbyshire, S32 5QW.
Tel: 01433 639777.
info@woolgathering.co.uk www.woolgathering.co.uk

Open 11am-4pm.

Entry Free

Accommodation YHA Hostel in village. B&B accommodation can be booked at Crown Cottage, which is a stone's throw from Woolgathering.

Yearly event or point of interest Derbyshire Open Studios – late May Bank Holiday weekend.

Directions Take the A623, and Eyam village is signposted southwest of Stoney Middleton.

Description Woolgathering is the studio for Deirdre Gage, feltmaker and textile artist as well as a gallery for emerging designer-makers. Courses are available on making felt by hand as well as basic design courses.

COURSES

Half-day course (for complete beginners) Felt-making – price includes laminated instructions and a starter felting kit.

1-day courses (for those new to felting) Felting Pictures – includes fibres and needle felting equipment; Felting Pots – includes fibres and threads.

2-day courses (for those new to designing) Inspired by verse, students bring their favourite poem or choose one from our list and design their own piece of textile art.

A Common Thread Same threads, same fabric, but different stitchers. See how diverse work comes from the same, common materials.

Inspired by Eyam Using the village, its features and history as inspiration for your work.

Design course dates by arrangement: please ring Woolgathering.

Woolgathering courses have a maximum of 4 people on a course. If you would like to attend a course with friends, please contact to arrange a time and content for your group (minimum 2 people).

HEREFORDSHIRE

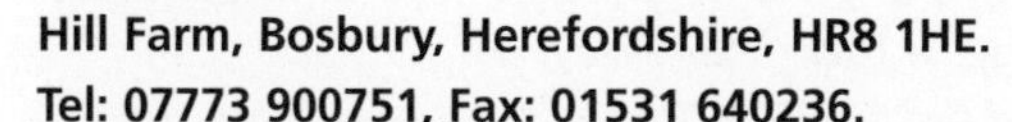

25. Childer Wood Heavy Horses

Hill Farm, Bosbury, Herefordshire, HR8 1HE.
Tel: 07773 900751, Fax: 01531 640236.
doug@heavyhorses.net www.heavyhorses.net

Accommodation B&B, camping.

Directions Sent to course participants on booking.

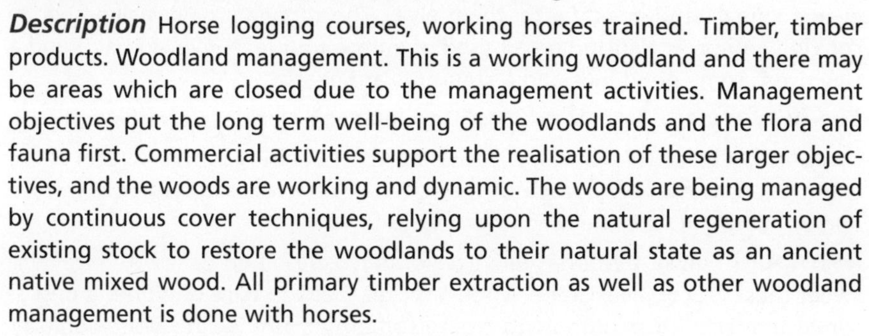

Description Horse logging courses, working horses trained. Timber, timber products. Woodland management. This is a working woodland and there may be areas which are closed due to the management activities. Management objectives put the long term well-being of the woodlands and the flora and fauna first. Commercial activities support the realisation of these larger objectives, and the woods are working and dynamic. The woods are being managed by continuous cover techniques, relying upon the natural regeneration of existing stock to restore the woodlands to their natural state as an ancient native mixed wood. All primary timber extraction as well as other woodland management is done with horses.

YOU CAN SEE (BY APPOINTMENT)

Timber-framed building • Sustainably produced timber • Wood fuel.

COURSES

Chainsaw Courses Fell trees that need to be felled as part of our ongoing woodland management; in particular the clearing of areas of Norway spruce and beech to allow natural regeneration of a mixed native woodland through continuous cover techniques.

Horse Logging courses Working horses and ponies offer an 'appropriate technology' for forest-friendly timber extraction. The ultimate low-impact 'base machine', they can be used at low capital cost to work with a range of equipment in environmentally sensitive sites or on steep or difficult terrain, working alongside or replacing tractors and heavy machinery. Full range of timber types and sizes extracted.

Woodland Management Courses Based upon our approach to profitable, small woodland management; theoretical inputs are based upon that ethos. Practical skills are developed doing the jobs that need doing, according to the time of the year.

26. Green Woodwork

Hill Farm, Stanley Hill, Bosbury, Ledbury, Herefordshire, HR8 1HE.
Tel: 01531 640125.
gudrun@greenwoodwork.co.uk www.greenwoodwork.co.uk

Accommodation and catering Camping pitches; basic greenwood huts for those without tents; a caravanning site on a neighbouring farm; or comfortable B&Bs and self-catering cottages a few miles away.

Other provision Running water, compost toilets, washroom and shower are all pleasingly integrated into the woodland setting. A hearty lunch is offered to those wishing to participate and contribute to a 'kitty' to cover costs. Tea, coffee and biscuits are provided at regular break times. For breakfast and evening meals students cater for themselves. The kitchen provides all necessary cooking equipment, including a raised open fire (our 'eye-level cooker') and a clay oven. Some good inns and restaurants are nearby.

Yearly event or point of interest Development Week at Clissett Wood.

Directions Clissett Wood lies five miles north of Ledbury in the Herefordshire countryside, within easy reach of Intercity rail services from London, Oxford and Birmingham, and the M5 motorway.

Description Clissett Wood is ten acres of mature ash and oak with hazel and sweet chestnut coppice. Green Woodwork teaches the skills of working unseasoned timber with handtools to create something unique for your home or garden. Gudrun Leitz offers two- to nine-day courses in Green Woodwork and Chairmaking for novices, amateurs and experienced craftspeople. In addition she makes a unique range of sculptural free-form furniture to commission. Workshop: There is a large well-equipped woodland workshop where there is a pole lathe, shaving horse and a set of tools for each course member. There are also steam-bending facilities, cleaving brakes and more.

YOU CAN DO

Volunteering. There are opportunities for skilled volunteers to assist on courses or contribute their cooking skills throughout the season. Food is provided and camping is free.

COURSES

3-day Greenwood Introduction A taster for beginners to become familiar with the basic tools and techniques of green woodwork. The course includes cleaving the wood with axes and wedges, and shaping it further with drawknives before turning it on the pole lathe; what to look for when buying tools; how to sharpen them; what woods are best to use, and more. Expect to leave with a stool and several turned items such as candlesticks, spoons, tool handles, baby rattles or projects of your choice.

6-day Green Woodworking and Chair Making Fundamentals After an introduction to green woodworking you can choose to make a high or low stool, a ladderback chair with bark seat, or a Windsor chair with solid elm seat. Alternatively you could make a pole lathe and shave horse or tackle several smaller projects.

Make and Use a Pole Lathe or Shave Horse Make your own pole lathe using a variety of carpentry and greenwood skills. Choose from one of the different 'models' suitable for the garden, garage or woodland or make your own shave horse and practise pole-lathing and green woodworking skills at the same time. For materials fee, see website.

Weekend: Rush Seating Course On this course Sheila Wynter tutors rush seating a ladderback or spindle-back chair. The course allows some makers on the nine-day Chairmaking Course to weave a really professional rush seat, but also has space for newcomers who should bring along the frame of a stool or chair to be rush-seated.

Wildwood chairs (6 days) This course teaches the combination of green woodworking and round wood techniques to make strong and individual items for home or garden. We examine the use and suitability of hazel and other coppice materials in combination with the traditional use of cleft and planked timber to make chairs, shelving, tables and garden seating. Students take home a very individual item of furniture.

LEICESTERSHIRE

27. Brocks Hill Country Park & Visitor Centre

      OD

Washbrook Lane, Oadby, Leicester LE2 5JJ.

Tel: 0116 2572888, Fax: 0116 2717356.

brockshill@oadby-wigston.gov.uk www.brockshill.co.uk

Open 10am-5pm Mon to Fri, 10am-4pm weekends and Bank Holidays.

Entry Free. The Park is open at all times. Access to 'behind the scenes' technology not currently available. Some events and activities require pre-booking.

Yearly event or point of interest Events held regularly throughout the year.

Directions Washbrook Lane is off the B582 Wigston to Oadby Road. Follow brown tourist signs.

Description Brocks Hill Country Park, Visitor Centre and café opened in 2001 after a successful grant application to the Millennium Commission. The visitor centre includes the UK's first ventilated solar array together with many other environmental features. The centre sits in a 67-acre country park with surfaced paths, 20kW wind turbine, human sundial and under 8s play area.

YOU CAN SEE (SOME BY APPOINTMENT)

Thermal mass • Specialist ventilation • Photovoltaics • Solar heating • Passive solar design • Wind turbine – medium (20kW) • Timber-framed building • Compost toilet • Rainwater collection.

YOU CAN DO

Volunteering • Programme of events for all ages.

CHILDREN'S ACTIVITIES

Holiday activities.

28. The EcoHouse, Groundwork Leicester & Leicestershire

OD

Parkfield, Western Park, Leicester, LE3 6HX.
Tel: 0116 222 0222, Fax: 0116 255 2343.
info@gwll.org.uk www.gwll.org.uk/ecohouse

Open Wed-Fri 2pm-5pm, Sat and Sun 10am-5pm.

Entry Garden, play area, café and shop – free. Showhome: adults £2, concs £1 (includes an audio tour). Guided tours for groups and hire of our Venue Room (capacity 35) need to be pre-booked. There are charges for both.

Yearly event or point of interest Annual Organic Plant Sale in May. Monthly Organic Gardening Course. See website for programme of monthly weekend events.

Directions The EcoHouse is on the A47 (Hinckley Road), Leicester. Entrance is signed 'Groundwork Leicester & Leicestershire', next to the gates to Western Park. **Bus:** Easily accessible by bus – ask for gates to Western Park. See website for up-to-date bus numbers.

Description The EcoHouse is Britain's original environmental showhome. Hundreds of environmentally friendly features are packed into the house and organic, permaculture garden to inspire visitors to make positive changes to their own homes. The EcoHouse is run by Groundwork Leicester & Leicestershire, formerly Environ, an independent charity and member of the national federation of Groundwork Trusts.

YOU CAN SEE

Photovoltaics • Solar heating • Wind turbine – medium • Renewable insulation • Compost toilet • Rainwater collection • Greywater system • Reed bed • Half of the building was built in the 1930s and has been retro-fitted with environmentally friendly features; half was a new eco-build in 2000 • Recycling area • Compost facilities on display, and advice on how to compost • Organic garden.

YOU CAN DO

Volunteering • Outdoor play area.

COURSES

Please check the website for the range of courses e.g. organic gardening, coppicing.

GREATER MANCHESTER

29. Bridge 5 Mill Centre for Sustainable Living

      OD MON-FRI

MERCi, Bridge 5 Mill, Centre for Sustainable Living, 22a Beswick St, Ancoats, Manchester, M4 7HR.

Tel: 0161 273 1736, Fax: 0161 274 4598.

merci@bridge-5.org www.merci.org.uk

Open Mon-Fri 10am-5pm.

Entry Free – donations welcome from individuals. Groups: fee for conference/ meetings depends on income of group. Please book for your visit, conference or tour of the building.

Yearly event or point of interest Contact us to arrange a tour of the building and learn about green refurbishment – Bridge 5 Mill has been refurbished with 95% reclaimed materials, and the building features innovative energy saving, reuse and rainwater capture systems.

Directions **On foot:** About 15 minutes from Piccadilly Train Station. At the bottom of the ramp on Piccadilly Approach, turn right up Ducie Street. Walk over the bridge, turn right and drop on to the Ashton Canal at Paradise Wharf. Turn left on to the canal tow path; continue past Piccadilly Village and under Great Ancoats Street. At bridge no 5 walk up the steps on to Beswick Street. Turn right and Bridge 5 Mill is the grey building on the opposite side of the bridge on your right. **Bus:** Bus no 216, near Nobel's Amusement Arcade, ask for Beswick Street. Get off just after the Mitchell Arms pub. Walk back towards the pub and take an immediate right, up Beswick Street. Continue up the hill and on the left, just before the Ashton canal bridge, you'll find the entrance to Bridge 5 Mill. **Car:** From the city centre: make for Great Ancoats Street (the inner relief road) east of Piccadilly train station. Travel up Old Mill Street, past the old Ancoat's Hospital on the right until you come to a set of traffic lights at the crossroads of Butler Street, Beswick Street and Bradford Road. Turn right down Beswick Street; immediately over the canal bridge is Bridge 5 Mill on the right. Parking – please park on Pollard Street or Beswick Street, although those with a disabled badge may park adjacent to the Centre. Parking is severely restricted during match days due to proximity of Manchester City football club.

Description The charity Manchester Environmental Research initiative (MERCi) owns and runs Bridge 5 Mill, Manchester's first Centre for Sustainable Living. It was refurbished with EU grants and the support of volunteers and trainees, and is probably Manchester 's most environmentally friendly building. Manchester Civic Society made an award to Bridge 5 Mill in 2001 as Manchester's most energy-efficient building.

This converted mill is home to a mix of charities, campaign groups and green businesses. The centre provides community resources, including meeting rooms, a library, an exhibition and a non-Microsoft computer suite. The building also has an excellent conference room available for hire, and tours of the building can be incorporated into your event. When funding is available a wide range of events are held, both in the mill and outside in the gardens, from green fairs to organic gardening; please phone for further details.

YOU CAN SEE

Straw-bale building • Cob building • Renewable insulation • Compost toilet • Rainwater collection.

30. Manchester Development Education Project

 OD TUES, WEDS, THURS

Laurel Cottage, c/o Manchester Metropolitan University (MMU),
799 Wilmslow Road, Manchester, M20 2RR.
Tel: 0161 921 8020, Fax: 0161 921 8010.
info@dep.org.uk www.dep.org.uk

Open Tues, Weds and Thurs from 12-5.30, or by appointment at other times.

Entry Free. Access to the Resource Centre is drop-in during our normal opening times times, or by appointment at other times. We can offer the use of a meeting room by arrangement, with access to Fair Trade refreshments.

Directions DEP is situated at the MMU Didsbury site. We are in Laurel Cottage, which is by the Sports Centre. **Bus**: a number of buses serve the site, including the 42, 142 and 157 from central Manchester. Didsbury is close to the M60 ring road junctions 1 and 2. **Train:** East Didsbury station is a short walk away.

Description Manchester DEP Teachers' Resource Centre and Bookshop has over two thousand teaching resources related to Global Education, ESD and participatory methodologies, selected to relate closely to teachers' needs. Visit our centre, or ask for our lists of resources on topics such as the Global Dimension, Environment, ESD, Literacy, Race, History and many more.

Manchester DEP also runs curriculum development projects and offers training courses designed to meet teachers' and schools' own needs, working with teachers, trainee teachers, youth workers and other educators to promote diversity, global education and sustainable development.

TEACHERS RESOURCE CENTRE

Examples of work in schools Training for teachers can be designed to fit their needs and their school's needs. Events can be run for pupils and year groups. Below are some examples:

• ESD / Sustainable schools • Simulation games for groups of varying size, up to whole school events such as an International Summit (see below) • Participatory learning methodologies such as Thinking Skills, Critical Literacy, Peer Education and Circle Time • Race equality • Citizenship • The Global Dimension to the National Curriculum • Teaching controversial issues.

An example of work with pupils: International Summit Simulation for Secondary Schools on a theme of global interdependence.

The simulation is a full-day event for a whole year group. Students take on the roles of countries, multinationals, Non-Government Organisations (NGOs), the media and the Summit Organisers or Secretariat. Each group researches and prepares in advance for the full-day event, which they organise and control themselves.

This simulation has been run successfully with pupils from year 8 to upper 6th. It has been run as an Earth Summit, a Commonwealth Heads of Government Meeting, and most recently as a model G8 Summit. It aims to help pupils understand the interdependence and interrelationships between countries, including:

Global inequalities • How resources are used and traded, and the need for trade justice • The process of international negotiations, including the role of multinational corporations, NGOs and the media.

The aim is to develop awareness of their roles as global citizens capable of contributing to positive change. Learning outcomes include developing skills of communication, debating, teamwork and co-operation; developing research skills using the internet; providing a stimulating educational experience for both students and staff, and illustrating the value to schools of visiting speakers offering different perspectives. And it's fun!

A full package from DEP to run a Summit simulation would include:

Preparation with the staff team • Session(s) with the pupils in the run up • Facilitating the event with the staff team.

NOTTINGHAMSHIRE

31. Attenborough Nature Centre

OD

Barton Lane, Attenborough, Nottingham, NG9 6DY.
Tel/Fax: 0115 972 1777.
enquiries@attenboroughnaturecentre.co.uk
www.attenboroughnaturecentre.co.uk

Open 7 days a week. Opening times times vary, but usually 10am-5pm.

Entry Free – some pre-booked activities carry a small charge. Please pre-book education / craft activities, holiday clubs and guided tours.

Yearly event or point of interest Spring and autumn provide the most interest for bird watchers.

Directions **Train**:15 mins walk from the train station. **Bus**: regular bus service (see website). **Foot/bike**: The Reserve and Centre are easily accessible on foot and by bicycle. Many visitors choose to arrive along the excellent riverside path routes (from either end) and then up Barton Lane. Cycling is permitted on these routes, but not on all paths around the reserve. **Car:** The Reserve is situated just off the A6005 between Beeston and Long Eaton. Turn on to Barton Lane off the A6005 at McDonalds, following the brown tourist signs. The car park is situated at the bottom of Barton Lane.

Description Attenborough Nature Centre, at Attenborough Nature Reserve, is an education and visitor facility developed and operated by Nottinghamshire Wildlife Trust, a registered charity with support from CEMEX and Broxtowe Borough Council. The centre provides both formal and informal education activities as well as high quality catering and conferencing facilities.

YOU CAN SEE

A range of renewable energy technologies including photovoltaic cells: heat recovery and solar panels are used to power and heat the building and a range of recycled and other sustainable materials were used in its construction.

Exchange heat pump • Photovoltaics • Solar heating • Passive solar design • Timber-framed building • Renewable insulation.

YOU CAN DO

Volunteering • Monthly nature walk.

COURSES

Various workshops are run throughout the year, such as 'willow lantern making' or 'paper making'. See website for more information.

32. Hockerton Housing Project

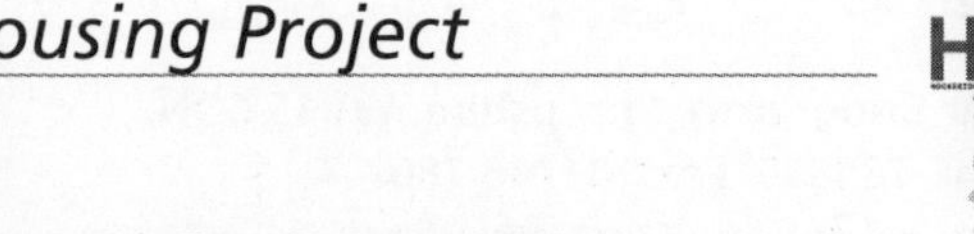

The Gables Drive, Hockerton, Southwell, NG25 0QU.
Tel: 01636 816902.
hhp@hockerton.demon.co.uk www.hockertonhousingproject.org.uk

Open All visits by the general public need to be pre-booked, but there is a lot of flexibility regarding times. Check the website for the range of tours available – these need to be pre-booked.

Directions Details are provided to all those with confirmed bookings on any tours, workshops or other events.

Description Since the completion of the houses in 1998, Hockerton Housing Project has established itself as an exemplar of sustainable development locally and nationally, providing a unique 'real-life' experience of living sustainably. This has resulted in the development of a range of services through the creation of a small on-site business, including guided tours, workshops, consultancy services, and school education visits. HHP runs a wide variety of events based at the Project, from basic tours to all-day technical workshops, from art events to venue hire. To find out which types of events would be most suitable for you or your group, or to customise your own visit, see the website. They also offer consultancy services.

YOU CAN SEE (BY APPOINTMENT)

Thermal mass • Specialist ventilation • Exchange heat pump • Photovoltaics • Solar heating • Passive solar design • Wind turbine – two 6kW (medium) • Compost toilet • Rainwater collection • Reed bed • The project is earth-sheltered, zero-heated, autonomous, zero-CO_2.

COURSES

Sustainable Community Workshop This event will be of particular interest to those interested in setting up, facilitating or joining a sustainable community. This includes individuals or groups with an interest in sustainable living/lifestyles, self-builders, landowners, and planners.

SHROPSHIRE

33. Dudmaston

  + FAMILY ACTIVITY ROOM

Quatt, nr Bridgnorth, Shropshire, WV15 6QN.
Tel: 01746 780866, Fax: 01746 780744.
dudmaston@nationaltrust.org.uk

Open 1st April-30th Sept.

House 2pm-5.30pm Tues, Wed, Sun.

Garden 12-6pm Mon, Tues, Wed, Sun.

Shop 1pm-5.30pm Tues, Wed, Sun.

Tea room 11.30am-5.30pm Mon, Tues, Wed, Sun.

Directions See website for details on cycling, walking, bus, train, road, ferry.

Description Late 17th-century mansion with art collection, lakeside garden and estate. See 'National Trust' in 'National Organisations' at back of book.

YOU CAN SEE

The house • Wood-energy scheme • Grounds.

COURSES

Please send a s.a.e. for information.

Fordhall Community Land Initiative

34. Fordhall Farm

     + PICNIC AREA

OD WED, FRI, SAT, SUN

Fordhall Farm, Tern Hill Road, Market Drayton, Shropshire, TF9 3PS.
Tel: 01630 638696.
project@fordhallfarm.com www.fordhallfarm.com

Open to the general public Wed, Fri, Sat and Sun 11am-4pm.

Entry Free (except for events). Please pre-book tours and courses.

Accommodation Eco-hostel and bunkhouse coming soon.

Yearly event or point of interest Family Fun Day every June, and the nature trail is open every weekend.

Directions Follow the A53 from Shrewsbury or Newcastle, and Fordhall Farm is about 20 mins from each, just on the outskirts of Market Drayton, next door to Muller Dairy UK. It is 20 mins from Junction 16 of the M6, or 40 mins from Junction 10a along the M54 and then A41 following signs to Whitchurch.

There is public transport from Shrewsbury, Newcastle, Hanley or Telford – just ask the driver to stop at Fordhall Farm.

Description Fordhall Farm has been chemical-free for over 65 years. It is a 140-acre pasture farm which has recently been placed into community ownership to save it from development. There are plans to open a tea room and an educational resource area. It currently has a nature trail, picnic area and farm shop. Fordhall is a work in progress and visitors can get involved with this. Short courses are run at Fordhall, and events are also held throughout the year. Ring the farm direct or see website for upcoming events.

YOU CAN SEE

The new buildings in 2007/8 will all be sustainably built and use renewable energy as far as possible. Exchange heat pump • Domestic wind turbine • Compost toilet • Rainwater collection.

35. Green Wood Centre (Small Woods Association)

OD EASTER UNTIL OCTOBER (EXCLUDING MONDAYS)

Station Rd, Coalbrookdale, Telford, TF8 7DR.

Tel: 01952 432769, Fax: 01952 433082.

marketing@greenwoodcentre.org.uk www.greenwoodcentre.org.uk

Open to the general public from Easter until the end of October, Tues-Sun (not Monday) 11am-5pm.

Entry Free access, but pay for coffee! Please pre-book tours and courses.

Accommodation Camping (hotel and B&B nearby).

Yearly event or point of interest Apple Day (2nd Saturday in October), Coracle Regatta on August Bank Holiday Monday.

Directions At western end of Ironbridge, Shropshire. See website for more details.

Description A visitor centre dedicated to the furtherance of wood, its growth, management, harvesting and uses, especially coppice management, and the use of wood as a sustainable building material. Demonstrations, trails, sculptures, interpretation, low-energy buildings, woodfuel district heating system, all on a beautiful wooded hillside in the Ironbridge Gorge World Heritage Site.

YOU CAN SEE

Biomass production – logs • Thermal mass • Specialist ventilation • Exchange heat pump • Solar heating • Passive solar design • Warmcell natural insulation • Timber-framed building • Compost toilet (soon) • Reed bed • and lots more.

COURSES

The Green Wood Centre runs around 60 courses a year teaching woodland management and coppice craft skills. Examples of courses are below:

Cider Making Turn windfalls into home-made cider. This day covers the history of cider, a description of apple types and sample tasting of basic varieties, milling and pressing apples (please bring along your own windfalls if you can). Students take away not only advice on fermentation to produce still and sparkling cider, but also a glass demi-john of juice.

Coracle Making Build an Ironbridge-style coracle using simple tools and techniques to suit all levels of experience and ability. The course includes: selecting suitable ash laths; techniques of bending laths to shape; nailing; clenching nails; fixing and tailoring the calico cover ready for painting.

Felt for bags Make a felt bag from a single piece of felt, working and moulding it using the traditional method of hot water, soap and friction to matt the fibres together. Students take home a unique bag made from the naturally beautiful colours of undyed fleece.

Fungi Foray This course looks at all aspects of fungi and ends with a fry-up of the day's best finds. Tutor John Hughes of Shropshire Wildlife Trust introduces you to stories surrounding fungi, folklore and religion, poisoners, and the vital role of fungi in the woodland ecology.

Hues from the Hedgerow This course looks at natural dyeing, using the plants and trees that grow in the woodland and hedgerow. It includes dying fleece, yarns or fabrics of your own choice in a variety of colours, learning how to experiment at home, and a chance to look at the use of woad and madder in the dyeing process.

Make a Cider Press Make a good-looking, robust cider or fruit press. This 10-litre press is based upon a traditional farmhouse design and will give approximately half a gallon of juice per pressing. The press is constructed from hard wood with steel press thread and stainless steel basket hoops. Workshop skills involve cutting, planing and jigsaw work.

Timber Framing Over a week the group constructs a small timber frame from green oak using traditional techniques. You will learn to scale and read working drawings, the principles of marking and cutting various mortice and tenon joints, the basic principles of roof construction, and the correct use and care of tools. Suitable for beginners and those who wish to develop further skills.

VTS (Vocational Training Scheme) OCN Introduction to Management Planning for Small Woods Aimed at those currently managing or planning to manage woodland, this course covers both practical and theoretical aspects of managing a small wood. The extended 2-day format allows for a more in-depth, applied approach that will include preparing an outline management plan and Woodland Grant Scheme Application for an area of local woodland. Jim Waterson is senior lecturer in rural affairs at Harper Adams University College

VTS Woodland Ecological Surveys Learn how to evaluate woodland and manage it for wildlife. This course is aimed particularly at woodland owners and managers, and includes evaluating the wildlife interest of woodland, determining priorities, and a range of management options to optimise the wildlife interest of an area of woodland. The course will be based at the Green Wood Centre and the surrounding Ironbridge Gorge Woodlands.

VTS Workshop skills: Rounding planes This is a foundation course in the use of rounding planes for beginners as well as those who are familiar with the tools but wish to further their knowledge of workshop skills. This 2-day course teaches the techniques required in the use of rounding planes for making stick and vernacular furniture such as Windsor and Shaker-style chairs.

VTS Workshop skills: Steam bending This 2-day course in steam bending includes selecting and steaming both seasoned and green timber for bending and shaping and learning the techniques required in the creation of vernacular furniture such as Windsor and Shaker chairs.

VTS Woodland Wildlife and the Law Held in Birmingham, this course consists of a morning lecture on the laws and regulations regarding wildlife and woodlands, and an afternoon spent in woodlands to see first-hand invasive plants and protected wildlife. Topics include: the current legal framework; legally protected species; invasive plants; designated sites such as SSSI; guidance in relation to biodiversity and woodland management; information on bats, birds, badgers and more.

36. Preston Montford Field Centre

  5 WORKROOMS, 3 LABS

Montford Bridge, Shrewsbury, Shropshire, SY4 1DX.
Tel: 0845 330 7378, Fax: 01743 851066.
enquiries.pm@field-studies-council.org www.field-studies-council.org

Open Not open to the general public.

Accommodation 120 beds in 2 residential blocks.

Directions **Train:** 5 miles from Shrewsbury station; taxi, or will pick up from station. **Car:** 1 mile from A5.

Description See 'Field Studies Centres' in 'National Organisations' section at back of book.

YOU CAN SEE

Wildlife garden • Weather station • Constructing a sensory garden.

COURSES

Beginner's Guide to Bumblebees This course introduces participants to the more commonly encountered bumblebees in gardens and the wider countryside. Emphasis is on field identification, but there is ample opportunity to study and identify prepared specimens using entomological keys and microscopes. Bumblebee ecology and conservation is explained, and a brief introduction to the other insects that make up this sub-order of *Hymenoptera*.

Beginner's Guide to Wild Flowers This course is for anyone who loves wild flowers and would like to be able to recognise and name more of them.

Bird Identification for Beginners During this weekend course you will learn about bird structure and the kind of observation skills which will help you to become proficient in bird identification.

Bird Survey Techniques A course for the keen amateur ornithologist or wildlife professional who would like to learn some basic bird survey techniques.

Dragonflies and Damselflies Shropshire's wealth of freshwater habitats boast a total of 29 different species of dragonfly, some of which are uncommon, visibly stunning and amazing to watch as they fly over ponds, rivers and upland flushes of the county.

Grass Identification This course uses a blend of laboratory-based identification workshops and field visits. It will help participants to identify grasses using field characteristics and botanical keys. Superb locations – ranging from limestone to acid, lowland to mountain environments – will be visited to ensure that a range of species are experienced.

Identification of Bats Rebecca introduces the world of bats, using slides, sounds and signs. Several roost sites and feeding areas will be visited, providing opportunities to see bats in their natural environment. There may even be some live rescue 'specimens'.

Land Mammal Identification An opportunity to encounter and identify many British mammals through field signs, live trapping and sightings. The course is practically based, with field visits and laboratory sessions designed to introduce a range of detection and identification skills. These include surveys for droppings and footprints, use of bat detectors, owl pellet analysis, identification of feeding signs, examination of museum specimens, small mammal trapping and evening watches for nocturnal mammals.

Moths and Late Summer Butterflies The course will cover aspects of butterfly and moth identification, trapping and information on the national notable species and life histories, and suggest methods to encourage these fascinating insects and their close relatives into our gardens. There will be daytime visits to a variety of habitats to observe late summer butterflies.

Songbird Identification The course will concentrate on developing identification skills by sight and song, using field notes and sketches as an important aid to observation. Late spring will ensure that there will be plenty of birds, whilst the comparative lack of leaves will hopefully mean that those which are heard can be seen as well.

Spider Identification Nearly 650 species of spider have been recorded from the British Isles. On this course we begin with a general introduction to spider biology, and collection and identification methods.

Trees and Tree Identification This weekend course is a general introduction to tree identification. By examining features such as tree shape, type of growth of both twigs and branches and studying buds, flowers, leaves and bark, participants will acquire the skills needed to name both British and foreign tree species.

Woodland Plants This course is for botanists wishing to further their skills in woodland plant identification, concentrating not just on trees and the more colourful woodland herbs, but also on the traditionally difficult groups such as woodland grasses, sedges and ferns. There will also be an introduction to mosses and liverworts.

Many further courses are available, including **Art, History, Tree Identification, Fungi, Safety, Mammal Identification, and more on plants, e.g. Identifying Difficult Lower Plants.**

STAFFORDSHIRE

37. Natural Sciences Centre

   OD MON-FRI AND SUN

Natural Sciences Centre, Newchapel Observatory, Stoke-on-Trent, Staffordshire, ST7 4PT.

Tel: 01782 785205.

newchapelobservatory@btinternet.com www.naturalsciences.co.uk

Open to the general public Mon-Fri and Sun, 10am-4.30pm.

Entry Adults £2, juniors and concessions £1. Please pre-book tours.

Directions The site is at the end of an alleyway off Newchapel High Street. It has a parking space for about 15 vehicles. There is a sign attached to a lamp post about 10 yards before the alleyway.

Description The Natural Science Centre is an educational and recreational visitor centre situated in Newchapel, one of the oldest villages in North Staffordshire. The Centre has a planetarium, observatory, alternative energy displays and a large conservation area. The Centre caters for all ages, and has wheelchair access. Guided tours and slide shows give a better insight into the many themes portrayed.

YOU CAN SEE

Solar heating • Passive solar design • Wind turbine – domestic • Timber-framed building • Rainwater collection • Recycling area • Composting • Observatory • Conservation area.

WORCESTERSHIRE

38. Abbotts Living Wood

Greenwood Cottage, Bishops Frome, Worcester WR6 5AS.
Tel: 01531 640005.
abbott@living-wood.co.uk www.living-wood.co.uk

Open Not open to the general public, but appointments can be made to visit the Centre for tours and courses.

Accommodation Camping pitches are available just inside the southern edge of the woodland with stunning views over the secluded Frome Valley. There are also three simple benders (hazel rod structures with a tarpaulin cover) for those who wish to sleep on-site without having to bring their own tent. Tucked into the woodland is a timber-framed compost toilet, a shower enclosed with a living willow and hazel screen, and a wash area with running water supplied from the farm bore-hole. For more conventional overnight stays, there are two B&Bs within a moderate walk/bike ride.

Yearly event or point of interest Open workshops during Herefordshire Art Week in early September.

Directions Between Bromyard and Bishops Frome in Herefordshire.

Description The aim of Abbotts Living Wood is to enable people to enhance their lives by harnessing two greatly neglected sources of green energy – trees and people. They run courses in green woodwork and chairmaking; give demonstrations and talks in greenwood skills; write books and articles on green wood matters, and make and sell wooden artefacts such as baby rattles and chairs.

The woodland facilities are completely free of mains electricity, powered almost entirely on wood harvested from the surrounding woodland, together with food mostly sourced from within the local area. The main structure is a covered open-sided workshop, purpose-built and fully equipped for up to eight students. Adjacent to the workshop is an area for cooking and eating, supplied with pots, pans, a multi-purpose wood-fired oven and an open cooking fire. Most people find they are amazed by the variety of food they can cook: as well as stews, curries and fry-ups, previous participants have indulged in home-made bread, pizzas, cakes, roasts, fruit crumbles and even a soufflé.

YOU CAN SEE (BY APPOINTMENT)

Biomass production – logs • Tile stove • Compost toilet.

COURSES

Living Wood courses Living Wood courses with Mike Abbott are held at Brookhouse Wood, just south of Bromyard. They run between May and

Compost toilet at Abbotts Living Wood.

Bath Organic Group take a break.

Looking for eggs at Soteriologic Garden.

A big bug hunt at The EcoHouse.

Cheesemaking by hand at Aldermoor.

Basket making at Brithdir Mawr.

Successful students at The Cherry Wood Project.

A fired bread oven made with clay from the forest at The Centre for Contemporary Art and Natural World.

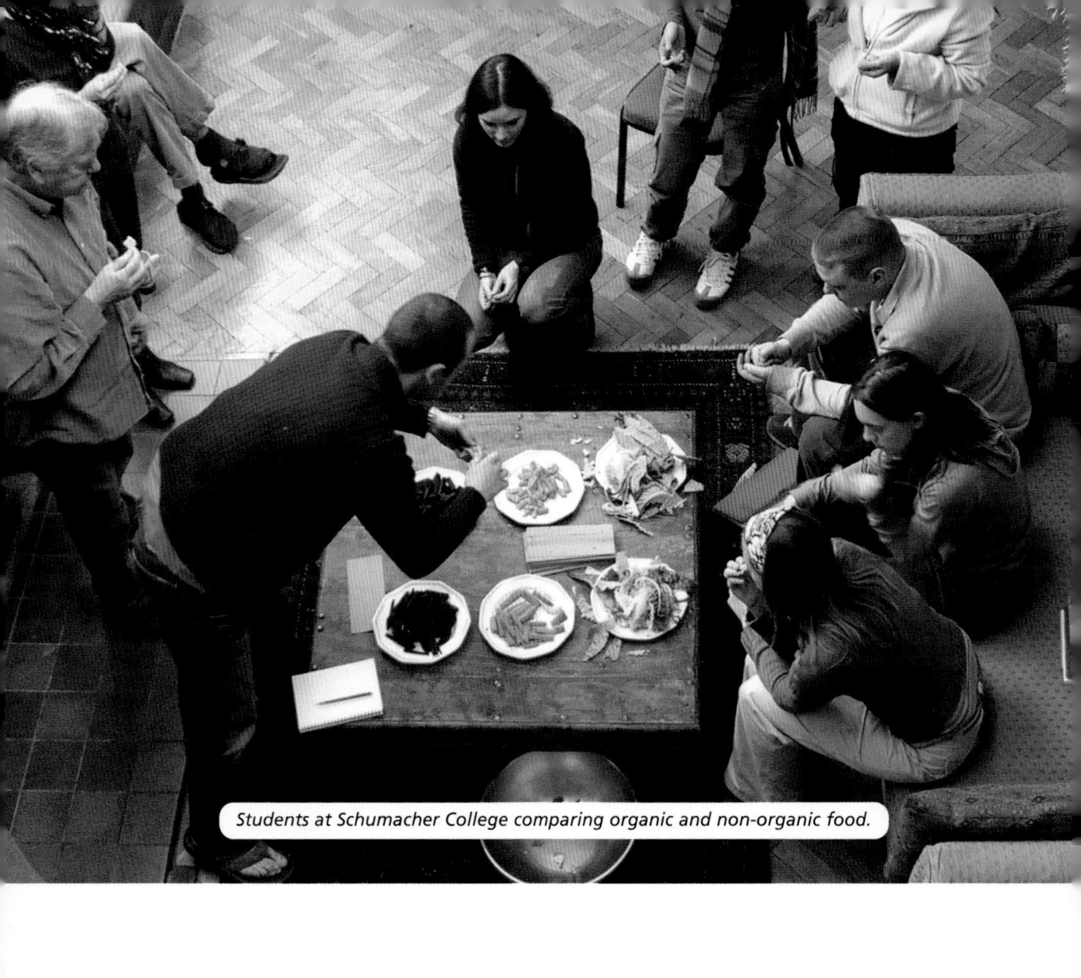
Students at Schumacher College comparing organic and non-organic food.

A willow workshop at Slack House Farm.

A healthy harvest at The Findhorn Foundation.

Working with Diamond the mare at Childer Wood.

Straw-bale students at Rippledown.

Yurt making at Future Roots.

An outing with Bristol Permaculture Group.

Ecos Millennium Environmental Centre.

Earthship Brighton – home of the Low Carbon Network.

Fashioning wood at Treewright.

Cleaving hazel at The Sustainability Centre.

Chair making at Abbotts Living Wood.

September and last for 3, 6 or 9 days. These courses have gained an international reputation. They are normally limited to seven participants, with two assistant tutors working alongside Mike. This level of individual attention means that all courses are suitable for all levels of experience from complete beginners to seasoned experts.

3-day Introductory Course Provides a useful taster covering the basic skills of cleaving, shaving and turning. These skills are then put to use in a wide range of projects such as candlesticks, spoons, baby rattles, hay rakes, tool handles, shaving horses, stools and more.

6-day Chairmaking Course Concentrates on the skills used in making a range of post and rail chairs, such as ladder-backs, spindle-backs and lath-backs. This includes all the basic green wood skills as well as steam-bending, seat-weaving, seasoning and chair assembly. Each student should finish the course with a completed chair capable of lasting a lifetime or more.

6-day Green Woodwork Course Gives the option of concentrating on chair-making or undertaking a range of projects covered in the introductory courses.

By combining a **3-day and a 6-day course** the student can make a chair as well as carrying out one or more of the smaller projects. Alternatively you could undertake a more elaborate chair such as an armchair or rocker. A substantial discount is offered for the combined courses.

In spring there are a number of 'development weeks' in which to get the workshop back into action after the winter hibernation and to carry out any new developments that are needed. This is an opportunity for anyone interested in setting up a workshop of their own.

3-week or 6-month assistantships Each year there is a limited number of places for longer-term training for those with some experience who might be interested in following a career in green woodwork.

What will you take away from a green woodwork course? Your own unique wooden product/s, created entirely from the log through your own efforts; the skills needed to continue on your own as a hobby or to develop further as a fulfilling occupation; an awareness of just how much you can achieve without the need for sophisticated powered equipment; probably the most lasting memory will come from the experience of working and relaxing with a group of like-minded people in a peaceful environment away from the pressures of everyday life.

Personal Tuition

Between November and March, Mike offers personal tuition for one or two people at a time in a purpose-built timber-framed workshop at his home in nearby Bishops Frome. This is the perfect way for students to concentrate on particular aspects of green woodwork and is suited to anyone wishing to take up green woodwork as a serious hobby or for a living.

Some possible topics:

Make and use a pole lathe	2 days
Making a shaving horse	2 days

Making a post and rung chair	5 days
Tool sharpening and modification	1 day
Advanced pole lathe turning	1 day

This is an idea of what can be covered, but each tutorial will be tailored to your precise needs and can include other topics of particular interest.

39. Bishops Wood Centre

Crossway Green, Stourport, Hereford and Worcester, DY13 9SE.
Tel: 01299 250513, Fax: 01299 250131.
www.bishopswoodcentre.org.uk

Open Pre-booked tours are open to groups.

Yearly event or point of interest Monthly open events run by Friends of Bishops Wood give opportunity for the public to visit – see details on website.

Accommodation Either local high quality B&Bs or camping at Bishops Wood, either in your own tent or a tipi.

Directions **Train**: The nearest railway stations are Kidderminster and Worcester Shrub Hill or Foregate Street. **Bus**: Well served by buses – see website for details. **Car**: From the A449 take the A4025 towards Stourport. At the top of the hill (400 yards) turn left into Bishops Wood Lane (signed Bishops Wood and Ombersley Golf Course). After approx. 100 yards turn left into the National Grid substation drive. Turn left into the car park. The Environmental Education Centre is on the left across the car park. Follow the yellow brick road to the round building over the bridge.

Description Bishops Wood is recognised nationally and internationally for its work in environmental and sustainability education with schools, and for its training courses.

YOU CAN SEE (BY APPOINTMENT OR AT EVENTS)

Specialist ventilation • Photovoltaics • Solar heating • Passive solar design • Wind turbine – domestic • Straw-bale building • Cob building • Timber-framed building • Renewable insulation • Compost toilet • Rainwater collection • Greywater system • Reed bed.

COURSES

The centre provides a stimulating environment for training, with indoor facilities and access to a number of habitats including oak woodland, meadows and a number of experimental environmental buildings. All the courses are aimed at environmental professionals or those with a deep interest in environmental issues. Most of the courses concentrate on how education and community development practice can help practitioners involve communities in sustainable development.

Advanced Forest Skills Complementing the introductory course, this course delivers a new syllabus of skills to empower group leaders with more confidence in leading groups in the outdoors. Includes fire-by-friction, wild foods, natural cordage and backcountry cooking.

Advanced Storytelling Workshop This weekend is playful and practical, focussing on the shaping and construction of stories.

An Introduction to Guided Walks This course equips those who have not led a basic guided walk with the skills to do so. Leaders with some experience will also benefit from reviewing their skills. The day will include: the whats and whys of guided walks, techniques, a practical exercise and more advanced techniques.

An Introduction to Leading Environmental Activities with Primary School Groups and How to Meet Teachers' Needs This day covers basic learning styles, leadership tips/techniques, accelerated learning and how to incorporate these into a learning programme. Practical activities are included to demonstrate how to motivate, organise and communicate with groups, and that all-important ingredient: facilitating full participation.

An Introduction to Storytelling Internationally renowned storyteller Ben Haggarty introduces people to storytelling techniques with a focus on working outdoors. Participants will come away with several stories to tell immediately and many ideas for finding and developing their own environmental material.

Creating Tangible Tales with Young People This day explores how to make environmental tales real for young people (from tots to teenagers) and looks at how to help develop sequential thinking, speaking and listening skills.

Creative Natural Adventures with Early Years Groups We look at ways of creating structures where wild ideas can grow and take shape, developing a sense of progression and change, and where our first stories can help us go on to explore the world around us. This workshop offers a range of simple activities to experiment with, ranging from quick outdoor adventures to drama, story-making and visual arts.

Drawing on Nature – environmental arts projects with schools and communities This new practical course involves drawing and sculpture using any material: wire, clay, charcoal, mud, grass, textiles and lots more.

Faces in Nature Working with puppets and performances in environmental projects with school and community groups.

Fieldcraft Skills Basic skills are taught in the context of providing a safe but stimulating outdoor classroom. These will include sensory awareness, understanding animal senses, techniques to improve our senses, observation trails, moving silently and unobserved, recognising animal tracks and signs, and elementary tracking.

Forest Skills Course This skills-based course is designed to provide the participants with practical know-how in the outdoors. Basic skills such as the safe use of cutting tools, firecraft and shelter building, are taught in the context of providing a safe but stimulating outdoor classroom. This course is particularly suited to those who use a woodland habitat to give children a meaningful experience of nature, such as forest school leaders.

Introduction to Environmental Youthwork (run in partnership with The Environmental Trainers Network) This day is aimed at environmental professionals, community development personnel and youth workers who are new to or have limited experience of the world of environmental youth work. We will examine what motivates young people outside the formal education system, and how to involve them in environmental activities and sustainable development initiatives.

Introductory Earth Education Workshop This annual workshop provides an introduction to this exciting, participative educational process. The day includes indoor and outdoor activities that will help learners improve their relationship with the planet, including an earthwalk, an 'energy flow' conceptual encounter and 'throwaway planet'. There will also be sessions on the philosophies behind earth education, leadership, and earth education programmes, including the newly released 'Rangers of the Earth'.

Passion for Place – Interpretation to inspire, delight and create This course is aimed at those who have had some experience of communication and/or interpretation, and are responsible for both planning and implementation of visitor services.

Public Events and Seasonal Themes A day on planning and inspiration for creative events with school and public groups. This day gives ideas for developing creative events, explores seasonal celebrations, looks at sources of inspiration, and explores a range of activities. We also address practical issues, risk assessments, health and safety, funding, and working with artists, storytellers and musicians.

Schools and Education for Sustainable Development (ESD) How organisations can work with schools to support them in delivering ESD.

Tales to Sustain A weekend where storytelling meets environmental education and activism.

Training the 'Environmental Trainer' This 1-day workshop examines the psychology of training. It will help those new to training and 'old hands' sharpen their skills and plan more effective training events.

Wild Words and Leafy Pages – literacy inspired by the natural world This workshop includes activities that can be used to help groups of all ages use language to explore, enjoy and celebrate their environment. Using spoken, chanted, sung and written words, it explores exciting ways of recording written words in bog books, wild scrolls, mapsticks, water pictures and hand-made paper.

Working with Living Willow This hands-on course looks at a number of planting and weaving techniques, where and how to source willow, and how to maintain sculptures and structures. Techniques will include looking at different types of wall, domes and tunnels; different types of play environment; constructing unusual sculptures, including figures; how to entice wildlife and combine a number of materials.

Working with Waste This workshop is aimed at recycling officers, teachers and other environmental educators working with waste. The day explores resources, activities, curriculum and the sharing of practice from various parts of the country. It looks at how sites – be they country parks or eco centres – and recycling officers can support and work with education groups, particularly schools.

40. Pump House Environment Centre

OD WED-SAT

The Duckworth Worcestershire Trust, Waterworks Road, Barbourne, Worcester, WR1 3EZ.

Tel: 01905 23267, Fax: 01905 734981.

info@dwt.org.uk www.dwt.org.uk

Open Wed-Sat, 10am-4pm.

Entry Free. Please pre-book tours etc.

Yearly event or point of interest The centre organises regular events and children's activities in partnership with the local authority.

Directions Follow the brown signs from the Barbourne Road (A38). Call The Pump House for further information, or see website.

Description The Pump House Environment Centre is situated next to the River Severn in Gheluvelt Park. The Victorian building was converted in 2003 to act as a community resource centre for the people of Worcestershire. The building demonstrates a number of sustainable technologies and building materials, including ground source heating and rainwater harvesting. Displays focus on energy, water and waste.

YOU CAN SEE

Exchange heat pump • Photovoltaics • Solar heating • Wind turbine – domestic • Renewable insulation • Rainwater collection • Wormeries • Real nappies

CHILDREN'S ACTIVITIES

Series of workshops held at the meadow site.

Key to symbols used

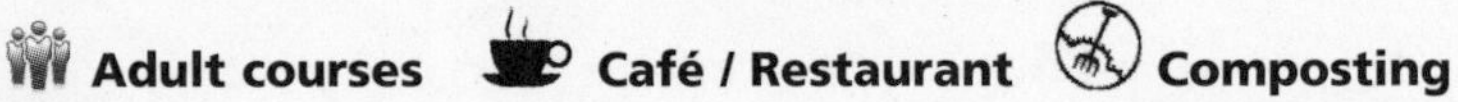 

Adult courses **Café / Restaurant** **Composting**

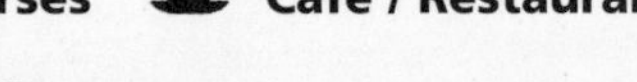

Garden **Grounds open to the public** **Meeting facilities**

 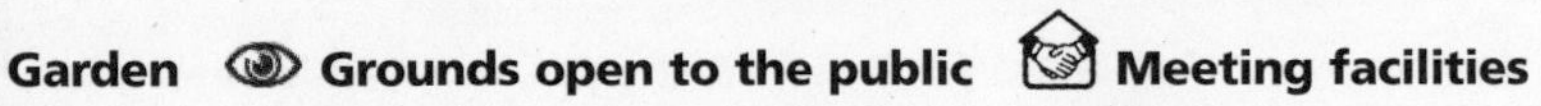

Recycling **School visits** **Shop** **Trails**

Visitor Centre **OD Open Daily (or on days specified)**

41. Wyre Forest Discovery Centre

OD + PICNIC AREA

Callow Hill, Bewdley, Worcs, DY14 9XQ.

Tel: 01299 266929, Fax: 01299 266302.

wyre.forest.discovery.centre@forestry.gsi.gov.uk www.forestry.gov.uk

Open Open daily (except Christmas Day) 10am-5pm (until dusk in winter). The use of the Centre needs to be booked for certain activities, so please check.

Entry Various, depending on group.

Directions Signposted off the A456 Kidderminster to Leominster road, 3 miles west of Bewdley at Callow Hill.

Description The Wyre Forest Discovery Centre is one of the Forestry Commission's education establishments. It runs programmes for schools and other groups, holiday activities for children, adult craft and wildlife focus events. All activities have an environmental theme, and use the forest as the teaching resource.

YOU CAN SEE (BY APPOINTMENT)

There will be new facilities soon, many of which will be accommodated in the new build.

YOU CAN DO

Volunteering • Walks and trails • Interpretation

COURSES

Craft days: Plant Supports, Hedgerow Baskets, Paper making, Felt making.

Wildlife Focus events: Deer Watching, Fungus Forays, Dusk Walks.

CHILDREN'S ACTIVITIES

Forest Friends – a club for pre-school children.

Various events throughout the year.

Key to symbols used

Adult courses **Café / Restaurant** **Composting**
Garden **Grounds open to the public** **Meeting facilities**
Recycling **School visits** **Shop** **Trails**
Visitor Centre **OD Open Daily (or on days specified)**

SOUTH-EAST & EAST ANGLIA

BEDFORDSHIRE

42. Marston Vale Services –
The Forest Centre

      OD

The Forest Centre, Station Road, Marston Moretaine, Bedford, MK43 0PR.
Tel: 01234 767037, Fax: 01234 762606.
www.marstonvale.org

Open 7 days a week, 10am-4pm.

Entry Free. Some things need to be pre-booked – please check.

Yearly event or point of interest Public tree planting each winter.

Woodworks Event: last year over 4,000 visitors watched professional craftspeople turn 'wood' into an array of products. Something for all the family, from green energy to horse logging.

Directions **Train:** Walking distance from Millbrook and Stewartby railway stations. **Car:** Off Junction 13 of the M1 onto A421 towards Bedford and follow the brown signs to the Forest Centre.

Description Visitor Centre set in 600 acres of woodlands and wetlands. The challenge is to use trees and woodland to transform 61 square miles between Bedford and Milton Keynes, repairing a landscape scarred by decades of clay extraction, brickmaking and landfill. Around 1 million trees have already been planted, and by 2030 there will be 5 million more by working with communities, government and businesses to create new woods and other wild places for everyone to enjoy.

YOU CAN SEE

Timber-framed building • Reed bed • Recycling area.

YOU CAN DO

Volunteering • Guided walks for wildlife • Practical conservation tasks • Various events during the year • Children's activities.

43. Randalls Farm Environmental Education Centre – The Wildlife Trust

OD

Broadmead Road, Stewartby, Bedford, MK43 9NE.
Tel: 01234 768542, Fax: 01234 765428.
randallsfarm@wildlifebcnp.org

Open 7 days.

Yearly event or point of interest Used for educational purposes, Watch and Greenwatch meetings, and holiday schemes.

Directions Just off the A421 between Bedford and Junction 13 of the M1.

Description A 6-hectare site in a semi-urban area. Placed between the land-fill site and the brickworks at Stewartby, Randalls Farm is an oasis for wildlife. The site is very easily accessed and used by a wide range of people. It is bisected by Elstow Brook. Habitats include ponds, woodland and a meadow.

YOU CAN SEE

Rainwater collection.

44. The Wildlife Trust for Bedfordshire, Cambridgeshire, Northamptonshire and Peterborough

OD

The Manor House, Broad Street, Great Cambourne, Cambs, CB23 6DH.
Tel: 01954 713500, Fax: 01954 710051.
cambridgeshire@wildlifebcnp.org www.wildlifebcnp.org

Open Reserves are open to the public all the time.

Yearly event or point of interest Has many events, so see website.

Directions See website for directions to individual reserves or offices.

Description The conservation charity works for a better future for all kinds of wildlife in the area. Its mission is to protect and improve habitats and wildlife while helping people to enjoy and understand their local wildlife. Many wildlife species and habitats have disappeared over the past 50 years. The Trust is working not just to protect what remains, but also to increase the numbers and diversity of native wild plants and animals in our countryside. It is not content with protecting what is left, as too much has already been lost – instead the charity wants to put something back.

The Trust runs approximately 130 local reserves and 6 education centres. Of particular interest are:

The Recycling and Waste Education Centre at Buckden Landfill Site • Forest Schools at Randalls Farm Environmental Education Centre • Grafham/Buckden Education Centre • Lings Wood Environmental Centre • Teacher training opportunities to support teaching about wildlife.

YOU CAN DO

Volunteering • Tours (need to be pre-booked) • Work parties for adults who want to learn conservation management skills and help enhance the nature reserve • Family events.

CHILDREN'S ACTIVITIES

Wildlife Watch is for children aged between 8 and 13 years who are interested in wildlife and the environment. Meetings are led by registered leaders, and are held on the first Monday of each month from 2pm-4pm.

COURSES (CHECK WEBSITE FOR CONFIRMATION)

There are courses under each of these headings:

Coppicing • Fungi identification • Habitat management • Hedgelaying • Plant identification • Sheep management • Teacher training • Understanding various vertebrates • Woodland products and crafts.

BERKSHIRE

45. The Living Rainforest

      OD

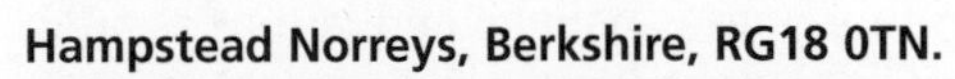

Hampstead Norreys, Berkshire, RG18 0TN.

Tel: 01635 202444, Fax: 01635 202440.

enquiries@livingrainforest.org www.livingrainforest.org

Open 7 days a week, 10am-5.15pm.

Entry £6.60, gives access to everything.

Directions **Train**: Mainline station is Newbury, with minibus link from bus station. **Car**: Follow the brown rainforest tourist signs from the M4/A34 intersection to the village of Hampstead Norreys. From Oxford, take the East Ilsley exit from the A34 and follow the signs.

Description The Living Rainforest is a live, growing rainforest under glass in the Berkshire countryside. On your visit, look out for free-roaming lizards, birds and butterflies, and discover plants that have changed the way we live today. In the new Millennium Commission-funded exhibition, 'The Human Impact', explore your impact on rainforests and other ecosystems, and learn more about renewable energies. More than half of the world's estimated 10 million species of plants, animals and insects live in the tropical rainforests.

Meet just a few of them live at the Living Rainforest, hear their amazing stories, see how they have adapted over millions of years.

YOU CAN SEE

Biomass boiler • Specialist ventilation • Photovoltaics • Passive solar design • Timber-framed building • Composting.

YOU CAN DO

Tours and group visits (over 15 people) need to be pre-booked • Talks for families.

46. TV Energy

Liberty House, New Greenham Park, Newbury, Berks, RG19 6HS.

Tel: 01635 817420, Fax: 01635 552779.

info@tvenergy.org www.tvenergy.org

Open Not open to the general public.

Yearly event or point of interest Annual conference. A range of workshops, seminars and site visits throughout the year. See the website for current events.

Directions Visit the website for location maps and directions.

Description TV Energy is a not-for-profit renewable energy agency for the Thames Valley, Surrey and north Hampshire. Its mission is to promote and facilitate practical sustainable energy solutions and provide education for communities, businesses, organisations and individuals within the Thames Valley and beyond.

YOU CAN SEE (BY APPOINTMENT)

Woodchip production • Biomass boiler • Biogas • Photovoltaics • Solar water heating • Passive solar design • Wind turbines – medium and large • Ground source heat pump • Energy crops (SRC) • Hydro • Earth building – cob.

COURSES

Courses and seminars in renewable energy for the public, farmers and businesses.

BUCKINGHAMSHIRE

47. Amersham & Wycombe College

Lycrome Road, Chesham, HP5 3LA. Tel: 01494 735555.
info@amersham.ac.uk www.amersham.ac.uk

Directions **Train**: London Underground serves Chesham Station, and there are regular bus connections. **Car**: Chesham Campus is adjacent to the A416 to Berkhamsted.

Description The College offers an education and training service to over 2,000 full-time and over 6,000 part-time students each year. It runs courses on three main campuses in south Buckinghamshire, in Amersham, Chesham and High Wycombe, and also at Cressex Training Centre in High Wycombe.

YOU CAN SEE (BY APPOINTMENT)

Exchange heat pump • Photovoltaics • Solar heating • Wind turbine – domestic • Renewable insulation • Rainwater collection.

COURSES

Renewable energy for the home An overview of various sustainable technologies that can realistically be installed in a home for those who want to know more about solar thermal, solar electric, wind power, underfloor heating and rainwater harvesting. Advice on how to calculate the financial payback period for the initial investment is included. The course enables you to make informed decisions about which technology best suits your circumstances.

Renewable Energy Systems Made Simple A range of new courses at our Environmental Training Centre, including:

Solar Thermal Installation for Installers For practising plumbers who wish to add solar thermal installations to their expertise.

Solar Electric Power Made Simple (Photovoltaics) For those who are interested in small-scale photovoltaics and are considering a non-grid connected installation. It covers location of the panels, design considerations, power calculations and applications, and advice on planning requirements.

Solar Thermal Energy Made Simple For those who are seriously interested in solar panels for their home and are considering a DIY installation. It covers location of the panels, design considerations, heat calculations and applications. Advice on planning requirements is also covered.

Wind Power Made Simple For those who are interested in small-scale wind power and are considering a DIY installation. It covers location of the turbine, design considerations, power calculations and applications and advice on planning requirements.

Other 'Made Simple' courses available: **Ground Source Heat Pumps, Rainwater Harvesting, Solar Thermal Energy, Underfloor Heating.**

48. Amersham Field Centre

FSC
BRINGING ENVIRONMENTAL UNDERSTANDING TO ALL

   2 WORKROOMS

Mop End, Amersham, Bucks, HP7 0QR.
Tel: 01494 721054, Fax: 01494 726893.
fsc@amershamfc.freeserve.co.uk www.field-studies-council.org

Open Not open to the general public

Directions Car: Off the A404 between Amersham and the M40.

Description The FSC manages the centre and nature reserve on behalf of National Grid Transco. See 'Field Studies Centres' in 'National Organisations' at back of book. Eco Centre Status

YOU CAN SEE

Wind turbine • Solar panel • Weather station.

COURSES

The Centre runs INSET courses throughout the year for both primary and secondary teachers. These are part of FSC professional development courses, or in partnership with Local Education Authorities. Contact the Centre for further information.

CHILDREN'S COURSES

The Centre offers a wide range of activities for children during holiday periods. These are aimed at children aged 5-7 and 8-11/14. Activities range from minibeast and pond-dipping, small mammal trapping and Native American days, to problem-solving, challenge activities and 'Harry Potter' days. If you would like to be included on our mailing list to receive regular updates about the programme of activities, please email the Head of Centre with your contact details.

49. Green Dimension

43 Clarence Road, Stony Stratford, Milton Keynes, MK11 1JE.
Tel: 01908 565250.
info@greendimension.co.uk www.greendimension.co.uk

Yearly event or point of interest The 1-day course 'Sustainable Energy for Homeowners' is ideal for homeowners interested in renewable energy options for their own homes. Online eco shop at www.click4eco.co.uk.

Directions Courses held at various locations – please contact for details.

Description Green Dimension provides engineering and environmental consultancy services, specialising in sustainable energy. It offers expertise in a range of renewable energy technologies such as solar thermal, photovoltaics, wind power, biomass, ground source heat and fuel cells. Green Dimension runs courses in sustainable energy for building and energy professionals and for the general public.

YOU CAN SEE

The location of the course will determine what there is to see. Most locations have some form of renewable energy to inspire students.

COURSES

To book courses see www.click4eco.co.uk.

Sustainable Energy for Homeowners (1 day) This course is designed to help homeowners and self-builders assess which sustainable energy options are most appropriate for their home. The course focuses on the main renewable energy options of wind, solar, wood heating and ground source heating. Participants will learn about which systems are likely to be most suitable for their home, as well as typical costs of energy generation.

50. Low-impact Living Initiative (LILI)

Redfield Community, Winslow, Bucks, MK18 3LZ.

Tel: 01296 714184.

lili@lowimpact.org www.lowimpact.org

Accommodation Eco hostel.

Directions **Train/bike**: There are two main cycle routes here from rail stations: Milton Keynes Central and Bletchley (which is a shorter ride) are both on the main London to Glasgow line; or Aylesbury is at the end of the Chiltern line from London Marylebone. **Train/bus**: train to Aylesbury, Milton Keynes or Oxford, then get a bus. **Car**: Redfield is on the A413 Aylesbury – Buckingham road, about 300m north of Winslow, on the right (next to a bus stop). See map. If it's dark, it's difficult to see, so go slowly or you could drive past it. Come straight down the drive and look for the signs for parking. Please park as far in as you can, so that the farmer can get past with his tractor. More travel details are on the website.

Description Low-impact Living Initiative (LILI) runs residential weekend courses at Redfield Community in Buckinghamshire. It also provides factsheets, books, natural products and home energy visits.

YOU CAN SEE (WHEN ON A COURSE OR BY APPOINTMENT)

Biofuel • Photovoltaics • Solar heating • Wind turbine – domestic • Straw-bale building • Earth building – cob • Timber-framed building • Renewable insulation • Compost toilet.

YOU CAN DO

Volunteering.

COURSES

Beekeeping • Build your own Geodesic Dome • Building with Timber • Cob Building • DIY for Beginners • Green Woodworking • Heating with Wood • Herbal Medicine • How to make Biodiesel • Living in Communities • Low-impact Smallholding • Make your own Essential Oils • Making and Installing Wooden Shingles • Natural Paints and Lime • Open Source Computing • Permaculture Gardening • Pruning and Care of Old Fruit Trees • Rammed Earth Building • Roofing • Round Wood Timber Framing • Self-build Solar Hot Water • Straw-Bale Building • Sustainable Energy for the Home • Sustainable Water and Sewage • Veg Oil Engine Conversion • Wind and Solar Electricity.

51. National Energy Foundation

The National Energy Foundation, Davy Avenue, Knowlhill, Milton Keynes, MK5 8NG. Tel: 01908 665555. info@nef.org.uk www.nef.org.uk

Directions **Bus/train**: Well served by bus and train links – see website for up-to-date info. **Car**: The National Energy Centre (NEC) is based in Milton Keynes, about two miles south of the city centre on Davy Avenue, just off H7 Chaffron Way, which runs parallel north of the A421.

Description NEF mobilises individuals, businesses and communities to make their contribution to reducing carbon emissions through energy efficiency and use of sustainable energy sources, in order to maintain the affordability of energy and to combat global warming. Its objective is to work for the more efficient, innovative, and safe use of energy and to increase the public awareness of energy in all its aspects. Currently it is working in the areas of renewable energy and energy efficiency.

YOU CAN SEE (BY APPOINTMENT)

Thermal Mass • Heat exchange pump • Photovoltaics • Solar heating • Passive solar design • Timber-framed building • Renewable insulation.

COURSES

1. Renewable Energy for Your Buildings (seminar) The National Energy Foundation is building on the successful provision of training over the past five years with a series of small renewable energy seminars. These will be held at the National Energy Centre in Milton Keynes. The seminars are designed to provide the background knowledge to the main renewable energy technologies, and information on grants and planning. They consist of a series of presentations and interactive tutorials followed by a tour of NEF's installations. The National Energy Foundation is also organising from time to time specific seminars in association with the trade associations. Subjects includes **Solar Thermal, Solar Electricity, Wood Fuel, Wood Pellets** and **Ground Source Heat**. See website for future events.

Who should attend and why? Renewable energy is a rapidly changing area, with new grants being introduced and government policy evolving. For those in the front line of local energy policy this is a unique opportunity to explore and expand their knowledge, with relation to all aspects of renewable energy. Briefing days are designed for Local Authority Officers, Architects, Designers, Planners, Energy Consultants and those who deal with the general public.

Agenda You already know that global climate change is one of the most serious problems facing us in the 21st century. You will also know at least a little about the potential benefits of renewable energy. So this seminar will be looking at the How and When, not just at the What and Why! How much solar resource is available in Britain? When should we choose one renewable source compared to another? How can we fund renewable energy installations? How can renewables help raise our profile and public image? When should we focus on energy efficiency, and when should we look at renewables?

2. Ground Source Heat Pump Seminar Sessions include: 'Real UK ground source heat pump installations – variations on a theme'; 'The European experience and how it relates to the UK'; 'Site conditions: Initial considerations before installing a ground source heat pump'; 'Geothermal drilling – It's happening now'; 'Heat pumps and underfloor heating – a marriage made in heaven?'; 'On the ground experience – the NEF case study'.

3. An event to suit your needs (can book for any centre) Renewable energy workshops and courses can be provided at venues around the country. These may be tailored to suit the needs and interests of individual organisations.

4. Home inspector – National Home energy rating training The most comprehensive short courses in energy efficiency in dwellings are probably those run in conjunction with the National Home Energy Rating scheme. Independent trainers run courses at various UK locations aimed at both the new build and existing homes sectors. From summer 2007, all English homes will be required to have a Home Inspection Pack (HIP) prepared before they are sold. This will contain legal searches, the Energy Performance Certificate required under the EU Energy Performance of Buildings Directive, and an optional Home Condition report. The Government estimate that there is a need for at least 4,000 qualified Home Inspectors.

52. The Environment Centre on Holywell Mead

    OD TUES, THURS, SAT, SUN

Bassetsbury Lane, High Wycombe, HP11 1QX.

Tel: 01494 511585.

manager@ecobuzz.org.uk www.ecobuzz.org.uk

Open Variable – please check website.

Entry Free. Facility hire needs booking.

Yearly event or point of interest Programme of exhibitions on environmental issues.

Directions Please consider taking the opportunity to walk, cycle or catch a bus when you visit. **Bus**: A40 London Road route – ask for the 'Cricket Ground' stop. **Train**: High Wycombe is the nearest train station. **Car**: From A40 London Road, turn into Bassetsbury Lane and follow signposts to The Rye and Holywell Mead Swimming Pool.

Description Since opening in November 2002, The Environment Centre has become a focal point for the community to investigate environmental issues. Providing an affordable meeting space for local environmental and community groups, exhibitions on a wide range of subjects, a base for field study groups, educational activities for children and adults, and volunteering opportunities, which are vital to the Centre's success.

YOU CAN SEE

Biomass boiler • Passive solar design • Renewable insulation • Recycling area • Composting area • Exhibitions.

YOU CAN DO

Volunteering • Allotment group • Walks with 'Simply walk'.

COURSES

Beekeeping (7 evenings) This is an evening course for beginners with tuition by skilled beekeepers. Classes include hands-on work with bees at our teaching apiary with equipment provided.

Master Composter Training (2 days) Master Composter Training is offered by Buckinghamshire County Council as a means of reducing waste and promoting home composting in the community.

Simply Walk (1 day) 'Simply Walk' is a scheme organised by the Primary Care Trusts and District Councils in southern Buckinghamshire, as part of the national 'Walking the way to health' initiative. Train to be a volunteer Walk Leader in this 1-day course.

53. The Parks Trust – Milton Keynes

 OD

1300 Silbury Boulevard, Milton Keynes, MK9 4AD.

Tel: 01908 233600.

info@theparkstrust.com www.theparkstrust.com

Open 24 hours a day.

Entry Free, but there is a charge for some events. Buying a school holiday pass will allow a child to access all events throughout the year for free.

Yearly event or point of interest The Parks Trust run over 200 events throughout the parks and offer environmental educational activities to schools, clubs and adult groups. See website for listing.

Directions There are no visitor centres at the present time. The Parks Trust manages 4500 acres of green space located throughout Milton Keynes. For more details about specific parks, see website.

Description The Parks Trust is the independent, self-funding charity that cares for the majority of Milton Keynes' parks and green spaces – the river valleys, ancient woodlands, lakesides, parks and landscaped areas alongside the main roads that make the city such a great place to live, work and visit.

YOU CAN DO

Volunteering • Women's walking network • Nature walks • Huge events listing on website.

CHILDREN'S ACTIVITIES

Tots in the park – list of activities • School holiday events • Activity sheets • Nature detectives.

COURSES / INFO DAYS

Try Coppicing An important woodland management technique – you can try your hand at coppicing.

Birds of a Feather A chance to make a bird feeder, learn more about our feathered friends, and build a bird box.

Try Fishing Call to book a half-hour fishing session at Teardrop Lakes with a member of the Angling Association.

Try Formative Pruning Introduction to one of the techniques of woodland management.

Try Hedge-Laying Try your hand at this traditional craft.

Try Helping Wildlife The Parks Trust provide more nesting boxes for local birdlife.

Try Meadow Restoration A chance to help the Parks Trust return an area of parkland known as Hills and Hollows to rich and colourful meadow.

Try Pond Management Learn more and lend a hand with creating better pond habitats for the wildlife within and around them.

Try Reed Cutting Reed beds need regular attention to ensure they provide the best possible habitat.

Try Scrub Management Try your hand at this management technique.

Also learn about ponds, nightlife, the importance of nettles, dragonflies, wildflowers, moths and birdwatching.

CAMBRIDGESHIRE

54. Houghton Mill

  HANDS-ON EXHIBITS, WORKING FLOUR MILL

Houghton, nr Huntingdon, Cambridgeshire, PE28 2AZ.
Tel: 01480 301494.

houghtonmill@nationaltrust.org.uk

www.nationaltrust.org.uk

Open See website, as times vary.

Accommodation Camping.

Directions Bus, cycling, road, train and walking directions on website.

Description Large 18th-century timber-built watermill – see 'National Trust' in 'National Organisations' at the back of the book. Guided tours.

YOU CAN SEE

Hydro-electric plant • Wildlife habitat • Grounds.

55. Wood Green Animal Shelters

    OD

London Road, Godmanchester, PE29 2NH.

Tel: 0870 190 4090.

info@woodgreen.org.uk www.woodgreen.org.uk

Open to the general public 7 days a week, 10am-4pm.

Entry Free (small charge for events). Some events need to be pre-booked.

Accommodation Camping.

Yearly event or point of interest There are numerous events held at Wood Green every year – for a full list, visit the website.

Directions Car: Leave the A14 at J24, join the A1198, follow signs for Royston. Follow for 1 mile and it is on the left.

Description The site of a full-sized wind turbine, this is the largest of their three Shelters, situated in 52 acres of Cambridgeshire countryside. It is home to a number of animals requiring new homes. The animals include dogs, cats, horses, goats, guinea pigs and rabbits; others are long-term residents such as sheep and llamas. A registered charity, Wood Green has been taking in lost and unwanted animals since 1924.

YOU CAN SEE

Wind turbine – full size.

ESSEX

56. Epping Forest Field Centre

  VISITOR CENTRE NEARBY CLASSROOMS

High Beach, Loughton, Essex, IG10 4AF.
Tel: 020 8502 8500, Fax: 020 8502 8502.
enquiries.ef@field-studies-council.org www.field-studies-council.org

Open Not open to the general public.

Accommodation Campsite nearby.

Directions **Car**: From the M25 (Junction 26) take the A121 towards Loughton, following the signs to the Epping Forest Centres. The Centre is located behind the Kings Oak pub in High Beach. From the A104, the Centre is signposted from both the Wake Arms and Robin Hood roundabouts.

Train: The nearest stations are Loughton (on the Central Line) or Chingford (from Liverpool Street). There are taxi services which operate from both stations. No bus travels to the Centre.

Description This purpose-built day centre was established by the City of London Corporation in 1970, the European Year of Conservation, and is managed by the FSC on behalf of the Corporation. See 'Field Studies Council' in 'National Organisations' at the back of the book.

YOU CAN SEE

Solar hot water • Eco centre • Water butts for garden and ponds.

COURSES

Discovering Badgers This flexible 1-day course on the natural history of badgers includes badger characteristics, anatomy and lifestyle. The course also looks at the badger in Essex – its position in the ecology of the area. Participants will have the opportunity to visit an active sett and gain tips on how to watch and track these beautiful creatures. Other features of the course include social structure, problems and hazards with badgers, and badgers and the law.

Discovering Bats This afternoon and evening course assumes no previous knowledge. It covers the biology of bats and their environmental requirements in general, with particular focus on British species. The course also covers the identification of bats in the hand and echolocation, and will provide a foundation for anyone wanting to train for a Roost Visitors Licence.

Family Environmental Arts Day This course is a fun introduction to using the outdoors as inspiration for creativity for all the family. Participants will make some pieces to take home and hopefully see the forest in a whole new way.

Green Man Arts Day Learn to use the forest as creative stimulus, and explore the folklore and traditions of the Green Man and the cycle of the seasons. Using natural materials and artistic methods, the spirit of the Forest is creatively explored.

Hedgelaying This 2-day course covers the history, ecology and management of hedgerows, and provides a sound knowledge of the techniques required for hedgelaying using traditional tools and materials.

How does your Garden Grow – an Introduction to the Science of Plants Students have the chance to look at the anatomy of plants and find out how they function and grow. The causes of common plant diseases are considered (do please bring along some problem material). The course ends with a session on improving your garden for wildlife.

Identifying Trees in Leaf September is an ideal time of year to study trees, as not only are they still in leaf but they also have their fruits to aid identification. Participants visit a number of sites in the forest, identifying many local species. Of particular interest to those who have a limited knowledge, the course is also suitable for those who know a little about trees and would like to know more.

Insects, Spiders and other Invertebrates This course gives an introduction to the often hidden world of insects and other invertebrates which surrounds us in woods, fields and gardens (and even in our houses!). It looks at how to examine, study and identify them and also looks at some of the literature which is available.

Involving Children in Environmental Activities This course is aimed at staff who want to involve children in exploring the environment. It investigates approaches to planning and delivering suitable activities. A number of techniques are examined including creative arts, earth education, educational games and activities in relation to the age range and formal education requirements of the group. It also examines the safety implications of working with children.

Mosses, Liverworts and Ferns The course covers the basic classification, identification, structure and function of mosses, liverworts and ferns, including microscope work. The programme is flexible, and suitable for both complete beginners and those with a working knowledge.

Natural Connections This innovative 1-day course helps to re-establish your connections with nature through physical and sensory exploration. The course focuses on the natural cycle of the forest, with particular emphasis on the seasons and traditions relating to them. Activities will include tasters of sensory exploration using the natural surroundings as stimuli, as well as small sessions of earth force meditation and awareness of the external internal. The course includes light physical movement and bodywork.

Natural History of Veteran Trees Well over 50,000 of the trees within Epping Forest are classified as veterans (trees of considerable age) which support a diverse array of organisms including birds, insects and fungi. This course offers the opportunity to view some of these special trees and looks at a selection of organisms which are associated with them. There is also the chance to discuss the problems of their future management.

Shieldbugs This course focuses on the characteristics that differentiate shieldbugs from other insects. Learn how to separate the four families and recognise some of the individual species, either as adults or nymphs (juveniles), and how to find them in the wild.

Survival for Beginners This 1-day course aimed at beginners gives participants a basic insight into fundamental survival techniques. It educates participants in the survival way of thinking by looking at nature as a renewable source, which if respected can be a survivor's best friend.

Using Environmental Crafts with Young People Course participants gain an insight into the opportunities for using environmental crafts as an educational tool as well as an encouragement for a variety of personal and social skills. They leave at the end of the day having had first-hand experience of a variety of crafts, and having acquired a basic knowledge of the properties of materials widely available.

Wild Food and Medicine This inspiring and informative 1-day course will identify seasonal foods: roots, aerial parts, leaves, twigs, flowers, fruits and seeds – all can be used to enhance flavour, to garnish and to vary one's diet.

There are many other courses listed on the website.

CHILDREN'S ACTIVITIES

Epping Forest Eco-Activities • The Centre offers a wide range of activities for children during holiday periods – see website.

57. Treewright

47 Eastbrook, Waltham Abbey, Essex, EN9 3AJ. Tel: 01992 711054.
Robin@treewright.co.uk www.treewright.co.uk

Open By appointment.

Yearly event or point of interest Workshop open day with green woodwork demonstrations and products for sale – see website for more details.

Directions Provided at time of making appointment or booking courses.

Description Treewright is a totally environmentally friendly business offering courses, demonstrations and talks about green woodwork, its uses and history, using only the pole lathe, shave-horse and hand tools to make interesting functional and fun furniture and treen (useful wooden objects) from freshly felled local timber. A wide range of products hand-made by Robin in our Waltham Abbey workshop is available all year round.

YOU CAN SEE – BY APPOINTMENT

Foot-powered pole lathes • Compost toilet.

COURSES

Green Woodwork Surprise yourself – make a whistle, spoon, spinning-top, or a stool. Transform a log into interesting useful treen to take home. Courses are by arrangement 1-5 days. Request a leaflet or see website.

HAMPSHIRE

58. Butser Ancient Farm

  OD

Nexus House, Gravel Hill, Waterlooville, Hampshire, PO8 0QE.
Tel: 023 9259 8838 www.butser.org.uk

Open Themed weekends and practical experiences only 10am-5pm (see website). Weekday visits can be pre-booked for groups.

Entry Adult £6, child £3.

Yearly event or point of interest Festival of Beltane in May.

Directions The farm is situated about 4 miles south of Petersfield in Hampshire, just off the A3. **Train/bus**: The nearest station is Petersfield, about 4 miles away. Bus No. 38 leaves from outside the station once an hour on Monday to Saturday. The walk from the bus stop to the farm is approximately half a mile (less if the driver will stop at Chalton Lane). **Car**: Travelling south from Petersfield, take the Chalton/Clanfield turnoff, and then almost immediately turn right. The farm is about 200 yards down on the right. Travelling north from Portsmouth, take the Chalton/Clanfield turnoff, turn right, right again at the mini-roundabout, then first left.

Description Butser Ancient Farm is a replica of the sort of farm which would have existed in the British Iron Age circa 300 BC. Founded in 1972, it moved to its present site at Bascomb Copse in 1991. The farm has buildings, structures, animals and crops of the kind that existed at that time. Much more than a museum, it is essentially a large open-air laboratory, where research into the Iron Age and Roman periods goes on using the methods and materials which were available at that time, and also by applying modern science to ancient problems.

YOU CAN SEE

Roundhouses • Ancient crop management • Ancient breeds of sheep • Roman Villa.

YOU CAN DO

The Activity Centre – gain hands-on experience of some of the domestic crafts and skills of the Iron Age and Roman periods. Visitors, both adults and children, are encouraged to grind grain on different types of quernstone, spin wool on a drop spindle, weave using a basic frame loom, plan a mosaic, make a clay pot and handle various artefacts.

COURSES

A series of practical experiences including:

Roundhouse construction Topics covered include the evidence and history of the roundhouse in Britain; materials and sourcing; carpentry in ancient times; use and design of tools; frame construction and engineering dynamics; wattle and daub (and alternatives); thatching methods and materials; living conditions and management. Includes practical experience.

Textiles Learn about evidence for clothing and materials; clothing styles in prehistory; use of fibres; loom design and use. Spin a yarn, dye with plants including woad; practise on a range of looms; weave patterns; look at construction of clothing.

Pottery *Session 1:* Evidence for pottery. Design and use of pots. Sourcing clay. Pugging and preparation. Pot construction. *Session 2:* Evidence for kilns and ways of firing. Construction methods of kilns. Firing pots using assorted methods (takes 24 hours minimum). We can help arrange overnight accommodation.

Flint knapping The course leader (Will) has been flint knapping for 30 years, and with his experience you will learn how to make a range of flint tools using coarse granite pebbles and deer antler hammers, taking home what you make. Flint has a number of interesting features, and a few surprises.

Herbs for healthcare in Roman Britain In making authentic medicinal recipes in the actual 'Roman Villa' at Butser, we can also explore how these herbs can help us today. Led by a medical herbalist, the course offers information on safe use of herbs in treating everyday problems. Stress is not new: the Romans were enlightened in psychology as well as surgery, and used fragrances to relax tensions. Mould rose petals with exotic ingredients to make the *rhodides* of Dioscorides, or weave herbs in wreaths to cure headaches. Learn to blend herbs to make pills, drinks and salves.

Hedgerow Basketry No previous experience is necessary to enjoy a day making a basket to take home. Learn about choosing the right stems, harvesting times and methods, storage and preparation of weavers and stakes. Practical experience in the ancient craft of basket weaving and design will lead to a working knowledge of the textures, appearance and flexibility of various natural materials.

There are other courses in ancient lifestyles and archaeology.

59. Greener Living

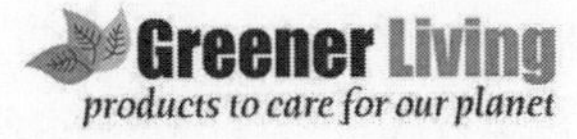

OD

327 Copnor Road, Portsmouth PO3 5EG. Tel: 023 9266 4700.

info@greenerliving.co.uk www.greenerliving.co.uk

Open 9am-5pm Mon-Fri, 10am-4pm Sat.

Directions **Bus:** The following buses stop within a few minutes walk of the shop; 1, 1a, 3, 5, 7, 7a, 24. **Train:** Alight at Portsmouth & Southsea. Call us for directions as it is a 2-mile walk from the station. **Car:** See website for directions.

Description A shop based in Portsmouth whose aim is "to offer people a choice and understanding of how to source alternative products which minimise environmental impact and sustain the long term future of their environment". The product range Includes natural paints made from clay and free from polluting biocides and toxic petrochemicals; natural carpets woven from sea grass, jute and coir; high quality wooden flooring manufactured from sustainable forests and treated with eco-friendly vegetable oils; building materials such as lime plaster, and Tereferno wall finishes; Thermafleece, sheep's wool insulation and Warmcell (recycled paper insulation); micro wind turbines, solar panels and heating systems that draw on sustainable resources; recycled stationery and paper, natural and energy efficient lighting, eco household products; and a full selection of eco baby products, such as nappies.

COURSES

Coming soon – see website.

60. Hampshire and Isle of Wight Wildlife Trust

OD

Beechcroft House, Vicarage Lane, Curdridge, Hampshire, SO32 2DP.
Tel: 01489 774400, Fax: 01489 774401.
feedback@hwt.org.uk www.hwt.org.uk

Open Please check with individual sites.

Yearly event or point of interest See website for details of over 350 events, activities and training opportunities every year.

Directions See website or contact for directions to individual reserves.

Description The Hampshire and Isle of Wight Wildlife Trust strives to create a better future for wildlife and wild places. It is the leading wildlife charity in Hampshire and the Isle of Wight, and is part of a nationwide network of 47 local charities. With the support of our 27,000 members, the Wildlife Trust is taking effective action to protect our natural heritage, helping it to flourish again. Find out more from the website.

FEATURES AT DIFFERENT SITES

Curdridge site Biomass boiler.

Testwood Lakes School visits, adult courses, shop, Southern Water dry garden, education centre.

Blashford Lakes School visits, adult courses, toilets flushed with rainwater.

Swanwick Lakes Centre Eco Centre, Visitor Centre, school visits, adult courses, recycling, composting, movement-sensitive lighting in toilet, timber-framed building, rainwater collection.

61. Herbs at Walnut

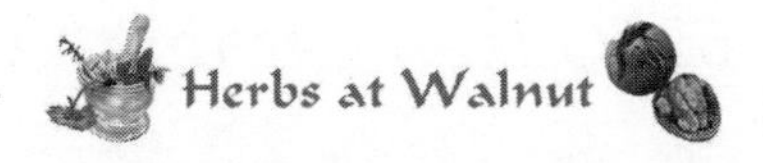

Walnut Cottage, Hampton Hill, Swanmore, Southampton, SO32 2QN.

Tel: 01489 891055.

info@herbcourses.co.uk www.herbcourses.co.uk

Accommodation Limited B&B sometimes available for those on courses.

Yearly event or point of interest Courses held on various dates between April and September each year.

Directions 3 miles from Corhampton (A32): follow signs to Swanmore via Park Lane into Hampton Hill. 11 miles from Winchester (M3 Jn9) via Morestead and Corhampton. Map and more detailed directions available on request.

Description Herbs at Walnut offers 1-day and 2-day courses on all aspects of using herbs for home remedies, making herbal remedies, herbal medicine, and using herbs in cooking.

COURSES

Choosing and growing herbs for home use (half day) Designed as a general introduction to growing and harvesting herbs at home. It covers the types of herbs to grow, which ones to gather from the wild, identification, drying and storage, designing a herb garden, and propagation.

Herbs for women (half day) Looking at various approaches to using herbs for treating women's ailments, drawn from both Eastern and Western healing traditions. The course includes a review of the properties of the more popular herbal remedies for gynaecological problems including black cohosh, sage, agnuscastus and Dang Gui.

Making and using herbal creams and ointments (one day) Combines practical sessions on making creams, infused oils and ointments with discussion on using the remedies in massage, herbal first aid and treating minor skin disorders and joint problems. At the end of the day, attendees will have a collection of at least four herbal products to take home.

The household medicine chest: making and using simple traditional remedies (one day) Focuses on the traditional use of herbs in both East and West for simple home remedies for treating household ills. There will be practical sessions on making remedies for internal use, including syrups and tinctures, and suggestions for treating everyday ailments such as common colds, coughs, and digestive upsets. Two herbal products to take home.

Using herbs for healthy eating and cookery (one day) A healthy digestion is the key to well-being. The course looks at herbal approaches to healthy eating, including Galenic and Chinese systems of food balancing. There will be sessions on using herbs in cooking, remedies for digestive problems, and the opportunity to make simple herbal oils and mustards. One herbal food product to take home.

62. Queen Elizabeth Country Park

CLASSROOM (HOLDS 250), THEATRE (HOLDS 42)

Gravel Hill, Horndean, Waterlooville, PO8 0QE. Tel: 023 9259 5040. www.hants.gov.uk/countryside/qecp

Open The Park is open daily. The visitor centre, shop and café: 10am-4.30pm daily 6th January to 28th February; 10am-5.30pm daily 1st March to 31st October;10am-4.30pm daily 1st November to 19th December.

Directions The Park is 4 miles south of Petersfield. **Train**: South West Trains Ltd run regular services to Petersfield. **Bus:** Bus services are operated by Stagecoach and First. Further information from Traveline (tel 0870 6082608). **Bike**: A cycle-way runs south from the Park, adjacent to the A3, towards Clanfield. **Car**: The site has its own sliproads on both the north and south bound carriageways of the A3. Follow the brown and white tourist board signs.

Description Hampshire's biggest country park with 20 miles of trails for walkers, cyclists and horse riders, Butser Hill National Nature Reserve and the highest point on the South Downs. The Park includes 1,400 acres of open access woodland and downland within the East Hampshire Area of Outstanding Natural Beauty. It holds a Green Flag Award, a Sustainable Business Award, ENCAMS accreditation as an Eco-Centre, and a Do it All Award 2006. Shop sells Hampshire Fare produce and 'QECP' local lamb and venison.

YOU CAN SEE

Reed-bed system • Local spinners demonstrate (check times) • Local arts • Local crafts • Butser Hill (organic status).

YOU CAN DO

Volunteering • 20 miles of trails for walkers, cyclists and horse riders • BBQ • Guided walk.

CHILDREN'S ACTIVITIES

Tots play trail • Over 8s play trail • Children's craft activities.

63. The Sustainability Centre

    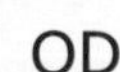 OD

The Sustainability Centre, Droxford Road, East Meon, Petersfield, Hampshire, GU32 1HR.

Tel: 01730 823166, Fax: 01730 823168.

Info@earthworks-trust.com www.earthworks-trust.com

Open 7 days a week, 10am-4pm.

Entry Free (donations are welcomed). Please pre-book tours, accommodation and room hire.

Accommodation 36 bed eco-hostel (Wetherdown), camping, yurt, tipi.

Yearly event or point of interest Yearly event in May focusing on fun and sustainability: children's events, renewable energy, demonstrations, care for the environment, traditional skills, a green forum and much more.

Directions **Train**: The nearest train station is Petersfield (about 10 miles away). **Bus**: Buses stop at Clanfield; then either get a taxi or walk about 2 miles to the Centre. **Car**: From the A3. Take the Clanfield exit (just south of Petersfield and Queen Elizabeth Country Park). Follow the brown tourist signs to 'The Sustainability Centre' (about 4 miles).

Description The Sustainability Centre, run by the charity Earthworks Trust, exists to provide visitors, the local community and volunteers with education, training and information on sustainability. It demonstrates to individuals from all walks of life and all abilities how they can develop their life skills and improve the quality of their lives in the context of sustainability, conservation and the environment.

YOU CAN SEE

Biomass production – woodchip • Biomass boiler • Photovoltaics • Solar heating • Triple glazing • Biodegradable flooring • Recycled paint • Timber-framed building • Renewable insulation • Compost toilet • Rainwater collection • Recycling area • Composting • Woodland Trail • Mobile phone self-guided trail • Animals • Forest garden • Herb garden.

COURSES

Apple Pressing and Bottling Day – Bring all Ye Windfalls Discover the classical seasonal apple harvest (early, mid and late varieties), gain detailed advice on localised varieties for planting, details of root stock, and possibilities of funding community orchards. Learn how to press and pasteurise apple juice, the different method of press and crush for fresh apple juice, and how to make home-made fresh cider the traditional way. Plenty of juice will be pressed for you to take home.

Autumn Herbs Enjoy an autumn herb walk to discover hips, haws and berries, their uses and folklore. Hunt for medicinal mushrooms, and learn their connection with plants and trees. Explore the uses and methods of making your own autumn syrups and liquors. Discover how to ward off winter ailments with warming and nourishing spices!

Beauty Without Costing the Earth Explore the art and science of beauty-potion making. You will create a variety of safe, 100% natural, organic, eco friendly skincare products, including cleansers, lip balms, moisturisers and babycare products using only natural, organic plant-based ingredients. A fun, hands-on workshop, where you will take home your own hand-made cream, recipes and other natural beauty tips.

Build your own Solar Thermal Energy Panel This course covers the theory, methods and techniques of building a working model solar water heating panel. With discussions, demonstrations and practical hands-on activities, you will gain a full understanding of solar water heating, its capabilities and limitations, and how to integrate it with an existing domestic hot water heating system.

Drawing Inspirations from Trees Using sight, taste, sound, smell, touch and intuition to explore the rich diversity found in trees, this course develops tree identification skills by comparing their colours, shapes and textures, plus leaves, fruit, seeds and bark. Exploring different creative and drawing techniques brings a love of trees sharply into focus.

Gate and Wattle Hurdles Learn to make a traditional chestnut gate hurdle and woven hazel hurdle. Using different techniques and tools, you will learn how to cleave poles and rods, shape and fashion chestnut poles and produce a solid wattle hurdle from the hazel rods.

Green Woodworking Select and fell a sycamore using an axe and saw. You will then select, cut and cleave appropriate material, carve and chop it into its final form. Learn techniques with traditional tools such as the froe, side axe, drawknife, shaving horse, whittling knife and stock knife and take home your own froe club, maul, spoons, spatulas and tent pegs, all made without using any power tools!

Hedgelaying Learn how to lay a local hedge 'Southern Counties Style' using traditional hand tools of bill hook, broad axe and hand saw. This highly practical course covers the history of hedgerows in our landscape, habitat importance, laying techniques, tool use, and care and safety.

Hedgerow Wines, Liqueurs and Cordials Gain practical experience in gathering fruits and herbs. Back inside learn the early stages of wine, liquors and cordials. Year-round recipes, advice on seasonal gathering and combining herbs safely. Plenty of samples to take home!

Herbal Christmas Gifts Prepare for Christmas by making your own natural gifts and decorations. Using natural materials, herbs and spices you can choose up to 5 of the following listed ideas costing between £3.50 and £7.50: The Yule Log, Snowman, Seasonal Seasoning Basket, Spicy Santa Sack, Pomander Beads, Spicy Star, Pressed Flowers Candle, Christmas Wreath, Cosmetic Herbal Basket and Miniature Garden. All materials provided.

Introduction To Biodynamics This day course introduces the guiding principles and methods of biodynamics; the history and importance of considering each setting – whether large farm or allotment garden; Learn how each setting exists in the field of polarity between the Earth and the Cosmos, and the influence of soil type, climate and aspect. Learn how to work with these factors and influence them through practical methods and biodynamic preparations.

Introduction To Permaculture Starting with the ethics and principles of permaculture, you will learn through theory and practical exercises how to apply permaculture to your life. The course introduces the design process, design tools and methodologies, and takes a detailed look at zones, sectors, and planning.

Lowering your Impact Using a simple 'footprint' technique, calculate your own resource use. Then by using creative activities the concept of sustainability is explored in order to discover what it really means to us. Armed with this knowledge, you will be able to apply this to your own household resource use.

Make Your Own Green Cleaning Products From small changes to complete change of routine, this course compares chemical-based cleaners with home-made. Learn how to make washing-up liquid, toilet cleaner and all-purpose cleaner to take home, and discover many other green tips for low cost, low impact healthy cleaning.

Natural Dyes for Wool and Silk A practical day identifying herbal dye plants. Explore the safe use and sourcing of mordants, how they work with dyes on different fabrics, partnering dyes with mordants correctly, and experimenting with each stage of dyeing on silk and wool. Take home a sample.

Permaculture in 5 Days For complete beginners and the more experienced, this course is centred on a series of practical design exercises, and considers the application of permaculture in the garden. Using talks, exercises and outdoor demonstrations, it will equip you to answer that all-important question: What's the best thing to do with my land?

Practical Coppicing Learn how to restore derelict hazel coppice and how to cut in-cycle coppice safely and efficiently. Using hand tools you will cut coppice, select, prepare and bundle coppice products, and discover the different products to be made. Learn tool techniques, care and sharpening methods and ecological and management aspects of coppicing.

Shingle Making Discover the art of shingle-making using traditional methods and only hand tools. Students will learn the importance of selecting suitable material; how to cleave oak and sweet chestnut shingles from rounds of wood with confidence; and how to dress the shingle using the side axe, draw knife and shaving horse.

Timber-Framed Building Learn medieval and modern framing techniques to gain a broad perspective and understanding of medieval box frame construction. Using traditional methods and hand tools, you will learn the basic skills of timber selection, techniques, maintenance and safety.

'Turning Into The Cycles Of Nature' Spring Herbs Take a herb walk with botanical and sensory exploration of local herbs using some Goethian methods to enhance your connection with plants and their environment. Learn how to make and use 'spring tonics' for detoxification and invigorating the body. Learn about folklore and tradition, the ritual of planting seeds, and how to take cuttings to develop your own herb garden.

Veggie Power Starting with an introduction in oil depletion and the need for alternative fuels, this course uncovers the technology behind vegetable oil powered diesel engines, along with the benefits of pure plant oil as compared with biodiesel. Enjoy practical discussions about fitting and removing a veggie power conversion kit, whilst getting under the bonnet of an already converted vehicle.

Wattle and Daub Learn the practical aspects of wall infilling using the age old practice of wattle and daub (sticks and mud!). Start by hand-cleaving sweet chestnut lathes and staves using the froe and 'A' frame to form the wattle weave. Finish the wall with the daub mixture, learning the technique, the ingredients and thermal insulation benefits.

Willow Christmas Craft Learn about willow varieties and their uses, coppicing, grading and dyeing willow, then how to harvest fresh willow using the Centre's own osier bed. Finish the day making your own hand-crafted willow items to take home.

Willow Craft in the Garden Learn to make willow lattice trellis, obelisks, bird feeders, hurdles and angels. Learn about willow varieties found in hedgerows, growing and coppicing osier beds, storage and soaking techniques. Go home with armfuls of willow goodies!

CHILDREN'S ACTIVITIES

Call for details of other courses and book beforehand – places are limited.

Fabric Fun Tie dyeing and printing with natural materials and stencils. Make your own simple print blocks to take home for cards and wrappings. Ages 5-11 years (8 years and under accompanied by an adult).

Hedgerows Discovery Trail Discover our hedgerows and edge-of-woodland trees, and the stories behind them. Collect elderflowers and make a cordial to take home. This is for parents and children to attend together (ages 6-11 years must be accompanied).

Pond Dipping and Fishy Crafts Discover the fascinating, slimy, creepy and beautiful creatures that lurk in our ponds. Practise some fishy crafts indoors, with items to take home. Ages 5-11 years (8 years and under accompanied by an adult).

Vikings How did the Vikings live? Play a Viking game to test your strength, make small Viking boats, watch Viking bread bake over the fire, invent a Viking saga, do some challenging work with hazel rods, and see soup bubbling in the cauldron. Ages 5-11 years (8 years and under accompanied by an adult).

Also **Survival Skills**

ISLE OF WIGHT

64. Aldermoor Earthworks

Aldermoor Farm, Upton Road, Ryde, Isle of Wight, PO33 3LA.

Tel: 01983 614795.

dave@aldermoor-farm.co.uk www.aldermoor-farm.co.uk

Open Please make an appointment.

Directions **Bus**: We support eco-friendly transport. From Ryde central bus station (next to the catamaran and hovercraft), take Bus 33 and ask for Upton Cross; walk 200 yards downhill to our entrance on the right. From Newport central bus station, take Bus 33, ask for Upton Cross; walk 200 yards downhill to our entrance on the right. **Car**: There is good car parking space. From Newport, follow signs for Ryde to Binstead, Havenstreet or on the Downs road. From Binstead, go straight through the traffic lights to the T-junction. Turn right on to Upton Road and drive to the newly laid hedge on the left: that's us.

Description Aldermoor Earthworks use their land and buildings to make workshops as practical as possible. They want to give people a chance to try things out, as well as get answers in their own situations. They feel that people get much more from guided activities than from lectures alone. So they ask presenters to include as much of the hands-on element of skills learning as possible.

YOU CAN SEE (BY APPOINTMENT)

Healing garden • Animal rehabilitation • Permaculture design.

YOU CAN DO

Volunteering.

COURSES

Healing for all • A Day of Falconry • Make your own Charcoal • Food from the Land • Green Inventors' Workshops • Cheese-Making by Hand • Maidens' Ceremony • Herbs and Healthcare • Introduction to Dowsing Earth Energies • Clearing Rituals on the Land • Hedgerow Wines • Food From the Land • Wand-making and Wisdom from Trees • Chemical-free Cleaning your Home • Carve a Green Man.

KENT

65. Canterbury Environmental Education Centre

     NATURE RESERVE

OD APRIL TO OCTOBER, SUNDAYS ONLY IN TERM-TIME

Broad Oak Road, Canterbury, Kent, CT2 7PX.

Tel: 01227 452447, Fax: 01227 456944.

canterbury.environmentalcentre@ kent.gov.uk www.naturegrid.org.uk

Open School Holidays: open Monday to Friday and Sundays 10am-4pm.

Entry £2 entry, under 5s free.

Yearly event or point of interest Family events throughout the year including at Easter, Halloween and Christmas.

Directions Located on Broad Oak Road in Canterbury, approximately halfway between Kingsmead Road and Vauxhall Road. The Centre entrance is next to a Mercedes showroom and opposite the HSS hire shop and Nursery and Toys.com (formerly D&A Toys).

Description Set on a beautiful 12-hectare site, run by Kent County Council and supported by National Grid, the main focus of the organisation is environmental education at primary level; there is provision for two classes a day of carefully tailored programmes. It offers a programme of family events, runs INSET days and hosts the Canterbury Urban Biodiversity Survey, thus enabling the wider community, volunteers and students to get involved with a variety of activities.

YOU CAN SEE

Recycling • Composting • Rainwater collection.

YOU CAN DO

Volunteering.

FAMILY ACTIVITIES

See website for up-to-date information.

66. Commonwork

Bore Place, Chiddingstone, Kent, TN8 7AR.

Tel: 01732 463255, Fax: 01732 740264.

info@commonwork.org www.commonwork.org

Accommodation Converted historic house and farm buildings.

Yearly event or point of interest Summer Public Open Day.

Directions Signposted, one mile north of the B2027 between Hildenborough and Edenbridge. Closest village is Chiddingstone Causeway.

Description Commonwork is a small group of charitable trusts and rural enterprises, founded in 1976 and based on an organic farm in the low Weald of Kent. Commonwork runs a conference and study centre for its own work and for hire, runs education programmes for schools and community groups, and seasonal events for the public.

YOU CAN SEE (BY APPOINTMENT)

Recycling • Compost • Brick-making.

YOU CAN DO

Volunteer • Walk the Field Trail.

COURSES

The Green Wood Courses The courses are largely based in the workshop, which is kitted out with all the necessary tools. Experienced tutors explain the safe use of tools and techniques, whether you are new to green woodwork, or wish to improve on skills you already have.

Green Wood Workshop (4 days) • Baskets and Willow • Charcoal Making • Coppicing and Felling • Green Wood Weekend • Hurdle Making • Living Willow Structures • Rustic Chairs/Garden Seats • Tools • Trug Making • Woodland Camp.

67. Rippledown Environmental Education Centre

Ripple Down House, Dover Road, Ringwould, Deal, CT14 8HE.

Tel: 01304 364854, Fax: 01304 364820.

office@rippledown.com www.rippledown.com

Accommodation Eco-hostel.

Directions Situated on the A258 between Dover and Deal, in Ringwould Village. For further information including public transport, see website.

Description Since 1977 Rippledown has provided students of all ages with opportunities for residential environmental education. During 2006 it has been running pilot courses: from straw-bale walls and lime render to working with rammed earth and cob. The programme of future courses will be based around sustainable building and rural skills training. See website for details of future courses.

YOU CAN SEE (BY APPOINTMENT)

Recycling area • Composting • Straw-bale wall on the 'Moot Hall' Activity Shelter. • Activity shelter – locally coppiced larch timber frame and an onduline roof • Low energy lighting where possible • Push-taps and 'hippo' bags • Rainwater collected in the gutters fed directly into pond.

COURSES

Cob oven • Lime render • Rammed earth • Straw-bale.

CHILDREN'S ACTIVITIES

Earthlings

A week of hands-on eco-investigation into the world around us: search the shoreline, peer into the pond, meet another mammal. We are all part of habitat Earth, so become a certified 'Earthling'. Suitable for children aged 8 to 12 with an interest in nature.

68. Romney Marsh Visitor Centre

  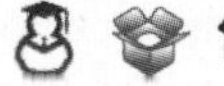
OD FRI-MON OCT-FEB;
7 DAYS MAR-SEPT

Dymchurch Road, New Romney, TN28 8AY.

Tel: 01797 369487.

rmvc@kentwildlife.org.uk www.kentwildlifetrust.org.uk

Open Fri-Mon Oct-Feb, every day Mar-Sept, 11am-4pm.

Entry Free.

Directions From New Romney follow the A259 for ½ mile and turn left opposite the golf course just before the Romney, Hythe & Dymchurch Railway bridge.

Description Managed by Kent Wildlife Trust, this eco-friendly building houses beautifully illustrated displays of the unique social and natural history of the Romney Marsh. The Centre is set in 11 hectares of country park created on land that includes Sites of Special Scientific Interest and is home to rare creatures such as the great crested newt, diving beetles and medicinal leeches.

YOU CAN SEE

Living roof (sedum) • Wood-burning stove • Straw-bale building • Timber-framed building • Rainwater collection.

LONDON

69. BRE (Building research and consultancy)

Bucknalls Lane, Garston, Watford, WD25 9XX.

Tel: 01923 664000, Fax: 01923 664010.

enquiries@bre.co.uk www.bre.co.uk

Open Not normally open to the general public, although training courses and special events are run providing access at different times.

Yearly event or point of interest BRE regularly runs events and training covering renewable energy in buildings and sustainable building design. BRE's Innovation Park is a demonstration of innovative modern methods of construction. Renewable technology can be accessed as part of courses/training activities.

Directions Close to the M1/M25 intersection between Watford and St Albans. **Train** The nearest mainline railway station is Watford Junction. Taxis are available at Watford Junction. Some car parking is available on site.

Description BRE is one of the world's leading centres of expertise for sustainable building, providing research, consultancy and training services.

WHAT YOU CAN SEE ON YOUR COURSE

Biomass production – woodchip / pellet • Biomass boiler • Biogas • Biofuel • Thermal mass • Specialist ventilation • Photovoltaics • Solar heating • Passive solar design • Wind turbine – domestic / medium / full size • Timber-framed building • Renewable insulation • Rainwater collection • Greywater system.

COURSES

BREEAM and EcoHomes Training courses on BRE's environmental assessment methodologies for buildings, covering homes, offices, schools and other buildings.

Building Better Schools • Home Inspector training • Part L and EBPD: training courses and events on new legislation and standards • **Planning for Daylight and Sunlight • Smart Homes Need Smart Controls • Low Carbon Technology Briefings • Biomass • Micro CHP • Photovoltaics • Small-scale Wind • Solar Thermal.**

70. Camley Street Natural Park

 OD THURSDAY TO SUNDAY, AND SEVEN DAYS A WEEK DURING SCHOOL HOLIDAYS

12 Camley Street, London, NW1 0PW.
Tel: 020 7833 2311.
www.wildlondon.org.uk

Open Thurs-Sun during term time and seven days a week during school holidays. 10am-5pm, or dusk if earlier.

Directions **Train:** Nearest station King's Cross St Pancras, King's Cross tube station and **Bus** routes 43, 63 and 214.

Description A unique wild green space in the heart of London, a stone's throw from King's Cross Station. This innovative reserve on the bank of the Regent's Canal was created from an old coal yard in 1984. It has a visitor centre and provides a natural habitat for birds, butterflies, amphibians and a rich variety of plant life.

71. Centre for Wildlife Gardening

    OD TUES, WEDS, THURS AND SUN

28 Marsden Road, London, SE15 4EE.

Tel: 020 7252 9186.

lwtwildgarden@cix.co.uk www.wildlondon.org.uk

Open Open to general public Tues, Weds, Thurs and Sun, 10.30am-4.30pm.

Entry Free. Appointments can be made for tours of renewable technology.

Yearly event or point of interest Activities for all the family on 'Frog Day', 2nd week in March every year.

Directions **Train:** Nearest stations East Dulwich and Peckham Rye. **Bus:** routes 37, 40, 176, 185, 484 and P4.

Description The Wildlife Trust's Centre for Wildlife Gardening is a must for anyone looking for ideas to attract more wildlife to their garden: take part in training courses, or just relax in the garden setting. Situated within a quiet residential street, the site has an award-winning visitor centre offering practical advice for city gardeners. Also a small nursery selling wildflowers, seeds, herbs, pond plants and native trees in season.

YOU CAN SEE (BY APPOINTMENT)

Passive solar design • Timber-framed building • Renewable insulation • Rainwater collection • Greywater system.

COURSES

Various 1-day courses for adults, such as:

Help House the Hedgehogs Learn to build hedgehog hibernating boxes.

Meadow Magic The last of our meadows; a heritage reserve.

Plant Remedies Explore Camley Street's plants and learn about some of the plant remedies.

Treemendous Fun Turn a keyring on the pole lathe; weave a basket; learn fascinating tree lore.

Wild Flowers and Butterflies Wild flowers and butterflies of Ealing's prime wildlife site.

Wild Flowers – Folklore and Herbal Remedies Learn about wild flowers and how to use them as natural remedies.

Woodland conservation Come and enjoy maintaining a secluded woodland reserve.

Other subjects include **Wildlife Gardening, Plant Propagation, Organic Gardening, Willow Weaving**.

CHILDREN'S ACTIVITIES

Creative recycling workshop Use recycled material to create wildlife creatures.

72. Permaculture Association (Britain)

BCM Permaculture Association, London, WC1N 3XX.

Tel: 0845 458 1805, Fax: 0113 2307461.

office@permaculture.org.uk www.permaculture.org.uk

Open Not open to general public.

Yearly event or point of interest The national Permaculture Convergence takes place every two summers – a networking gathering of hundreds of permaculture enthusiasts.

Directions Courses held at various locations (please contact for details).

Description Permaculture is an ecological design process that functions to benefit life in all forms. It works because it can inspire and empower you to create your own solutions to local and global problems, enhancing quality of life using local resources and your own skills and talents. The Permaculture Association supports the UK network through education and as a resource database.

COURSES

Permaculture Introduction Courses

They recommend that individuals who wish to know more about permaculture first attend a 2-day introductory course. This 'taster' will provide an opportunity to actively learn about the ethics and principles, and see examples of designs being implemented in this country and abroad. The course also provides the opportunity to discuss issues with other people and gain an overview of the growing permaculture network.

Permaculture Design Courses

These come in a variety of formats, and are offered as two-week residential courses, evening classes, a series of weekends, or a combination. They lead to recognition of the individual as a Permaculture Design Course Graduate. The Association promotes the design course as a vital stage in the development of understanding of permaculture ethics, principles, design processes and implementation techniques. The Permaculture Design Course has been accredited by the Open College Network. There are two courses: 90 hours and 120 hours, plus variety of shorter courses and courses in related subjects that are accredited. Accredited courses are often run with the Workers Education Association, and are offered at very low cost to those on low/no incomes.

The Diploma in Applied Permaculture Design and the Diploma WorkNet

This particularly applies to individuals who, as a result of completing the design course, are developing and running permaculture projects and want to go on to become teachers. The Diploma runs over a minimum of two years, and is based on an 'action learning' approach – learning by putting into practice. The Diploma recognises and celebrates the students' active work in applying the ethics, principles, design strategies and operational techniques of permaculture, and their achievements in developing practical sustainable systems.

73. South London Permaculture Soteriologic Garden

PO Box 24991, London, SE23 3YT.

Tel: 0845 458 1734.

Know@soteriologicgarden.net www.soteriologicgarden.net

Open Please call to arrange visits: times and days are flexible. Tours and events need to be pre-booked.

Entry £2.

Yearly event or point of interest August Bank Holiday Monday Green Man Fayre 12-6pm. One Tree Hill allotments.

Directions Come off the South Circular at Forest Hill, and head for Honor Oak Park train station. The allotments are behind the train station on One Tree Hill. Call first to view.

Description South London Permaculture run a number of different projects. Its hub of community activity is centred around the permaculture woodland project at the allotments. There one can see cob and green timber building. Various events throughout the year include a Green Man Fayre and Apple Day. There is also a mobile 16ft yurt project entitled Re-LEAF (Learning, Entertainment, Art, and Food).

YOU CAN SEE (BY APPOINTMENT)

Thermal mass • Cob building • Mud-block building • Timber-framed building.

COURSES

Fruit tree grafting (1 day) Get to know the whys and whats of successful tree grafting, and walk away with a tree. Techniques include whip and tongue, crown grafting and budding.

Introductory to Permaculture (2 days) A very hands-on introduction to understanding permaculture, with practicals and local walks taking in conservation, natural building and guerilla gardening.

Natural building (2/3 days) Put theory into practice using cob, green timber and other traditional methods, and see on-site structures in place. Covers anything from greening your roof to outdoor sculpture.

Permaculture 72-hour design course (8 alternate weekends) Permaculture in an urban environment. See how to apply it to the garden, balcony, house and business. An extensive look at the environment in a picturesque locality.

74. The Millennium Centre

        OD 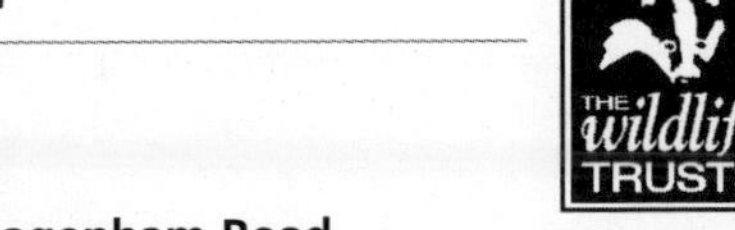

The Millennium Centre, The Chase, Off Dagenham Road, Rush Green, Romford, RM7 0SS.
Tel: 020 8593 8096, Fax: 020 8984 9488.
wildlifetrust@lbbd.gov.uk www.wildlondon.org.uk

Open Open to the general public 7 days a week, 10am-5pm.

Entry Free.

Yearly event or point of interest Events are organised on a monthly basis.

Directions Train/foot: Take Dagenham East District Line to Dagenham East. Cross over zebra crossing, then turn left over bridge. Walk past Aventis Pharma buildings. Turn immediately right along footpath, following factory boundary. Continue past the sports field on your left, until reaching a footpath to the left. Follow footpath into the country park. For The Millennium Centre, follow the path straight on past the woodland, and you will see The Millennium Centre over to the right.

Description Spend time exploring the Trust's largest reserve, incorporating shallow wetlands, reed beds, horse-grazed pasture, scrub and woodland. There is an impressive range of animals and plants, including the rare black poplar tree and spiny restharrow.

YOU CAN SEE

Specialist ventilation • Passive solar design • Wind turbine – full-size • Timber-framed building • Renewable insulation • Rainwater collection.

NORFOLK

75. Brancaster Millennium Activity Centre

Dial House, Brancaster Staithe, King's Lynn, PE31 8BW.
Tel: 01485 210719.
brancaster@nationaltrust.org.uk
www.nationaltrust.org.uk

Open Please phone for details.

Directions See website.

Accommodation Yes.

Description Large area of coastal habitat, particularly noted for birdlife.

See 'National Trust' in 'National Organisations' at back of book.

YOU CAN SEE

Solar panel • Wind turbine • Heat exchanger • Recycling • Photovoltaic and thermal solar energy systems • Grounds.

COURSES

Please contact or see website for list of courses and events.

76. Greenhouse Trust

     OD TUES-FRI

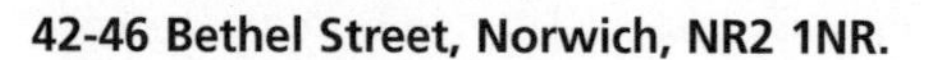

42-46 Bethel Street, Norwich, NR2 1NR.

Tel: 01603 631007.

info@GreenhouseTrust.co.uk www.GreenhouseTrust.co.uk

Open Open to the general public Tues-Fri (also Mondays in Dec) 10am-5pm. Please pre-book tours and use of the meeting rooms.

Yearly event or point of interest Greenhouse Trust aims to be an exemplar of sustainability, and provide information.

Directions Go towards St Giles from City Hall, along Bethel Street; The Greenhouse is just past the Police and Fire Stations.

Description The Greenhouse is a Grade II listed building which has been restored and refurbished as an example of sustainable living. It uses solar PVs and water tubes, harvests rainwater, and is insulated to very high standards using natural materials. It is sheltered by a PV roof. The Greenhouse operates an organic, vegetarian, fairtrade café and shop, using locally-sourced materials wherever possible.

YOU CAN SEE (BY APPOINTMENT)

Photovoltaics • Solar heating • Passive solar design • High levels of insulation using natural materials (cork, Warmcel and Thermafleece) • Rainwater collection • Building layout minimises energy use • Use of non-oil-based paints.

77. Norfolk Wildlife Trust – Cley Marshes

NORFOLK WILDLIFE TRUST

Bewick House, 22 Thorpe Road, Norwich, NR1 1RY.

Tel: 01603 625540, Fax: 01603 598300.

admin@norfolkwildlifetrust.org.uk www.norfolkwildlifetrust.org.uk

Open Please check website.

Directions NWT Cley Marshes is located on the north Norfolk coast, between Blakeney and Salthouse, on the A149 coast road.

Description Norfolk Wildlife Trust was established in 1926, and is the oldest of the 47 Wildlife Trusts in the UK. NWT Cley Marshes is the oldest nature reserve in the country, purchased by Dr Sydney Long with the express purpose of benefitting Cley's resident and visiting birds.

YOU CAN SEE

Exchange heat pump • Solar heating • Wind turbine – medium • Rainwater harvesting.

78. Sheringham Park Visitor Centre

Visitor Centre, Wood Farm, Upper Sheringham, NR26 8TL.

Tel: 01263 820550.

sheringhampark@nationaltrust.org.uk

www.nationaltrust.org.uk

Open See website.

Directions See website.

Description See 'National Trust' in 'National Organisations' in back of book.

YOU CAN SEE

Wood energy scheme.

YOU CAN DO

Guided walks.

COURSES

Bat and Moth Night Take an evening stroll with our warden to learn about the species of bats and moths in the park when they are most active.

Bat ID Workshop Learn more about how to identify different types of bats in this more advanced workshop. Refreshments included.

Bird Migration A slide show and talk on 'A year in the Life of a Bird'. Why do they migrate?

Dawn Chorus Take a morning stroll with local ornithologist Moss Taylor to hear and see the birds of Sheringham Park at their best. Coffee and croissant included.

Jewellery Workshop Learn the basics of jewellery making from 'found materials' as great gifts for your mum, sister or aunty. With Amelia Vargo.

Norfolk Smallholders Show and Market Come and meet the Norfolk Smallholders Training Group hosting their poultry show, with an extensive market of home-produced Norfolk goodies. Other livestock on display, trade stands and much more!

Other courses Please contact the Centre.

79. Wildfowl & Wetlands Trust, Welney Wetland Centre

+ BINOCULAR HIRE

Hundred Foot Bank, Welney, Wisbech, PE14 9HE.

Tel: 01353 860711, Fax: 01353 863524.

info.welney@wwt.org.uk www.wwt.org.uk

Open Open daily except Christmas Day. Mar-Oct daily 10am-5pm; Nov-Feb Mon and Tues 10am-5pm, Wed-Sun 10am-8pm.

Entry WWT Members free. Admission (excludes some special events and activities – group discounts* for 15+ people):

Adults	£4.50	*group rate £3.70
Seniors (60+)	£3.70	*group rate £2.90
Children 5-16	£2.70	*group rate £2.20

Yearly event or point of interest Celebrate the Swans' Return on 11 and 12 November. See thousands of winter-visiting Whooper and Bewick's Swans, and watch them feed by floodlight. Some items need pre-booking – phone or see website for details.

Directions WWT Welney is located 12 miles north of Ely, 26 miles north of Cambridge and 33 miles east of Peterborough. There is tourist signage from the A1101 in the vicinity of Welney village and from the A10.

Description New, specially designed, eco-friendly, two-storey visitor centre and landscaped grounds with improved facilities and increased space, including a heated observatory and hides. A fantastic base from which visitors can discover, explore and learn about the wealth of wildlife that exists on the WWT Welney reserve, including breeding birds in spring, dragonflies, butterflies and wild flowers in summer, and thousands of swans, ducks and other water birds in winter.

The Welney Wetland Centre is one of nine UK sites managed by the Wildfowl & Wetlands Trust – the largest UK specialist wetland conservation organisation.

YOU CAN SEE

Specialist ventilation • Exchange heat pump • Timber-framed building • Renewable insulation • Rainwater collection • Greywater system • Reed bed • Sustainable urban drainage system.

YOU CAN DO

Pond dipping.

COURSES / EVENTS

Many events and courses that include the whole family.

CHILDREN'S ACTIVITIES

Junior Bird Club (Ages 10-16) – Children can go behind the scenes with the Welney Wardens and improve their birding skills. Annual fee includes 4 sessions at WWT Welney (Sundays 10am-noon) and 2 off-site visits to other nature reserves.

OXFORDSHIRE

80. Sutton Courtenay Environmental Education Centre – The Berks, Bucks and Oxon Wildlife Trust

Berkshire
Buckinghamshire
Oxfordshire

Sutton Courtenay Road, Abingdon, OX14 4TE.
Tel: 01235 862024, Fax: 01235 862024.
SCEEC@bbowt.org.uk

Open Tue-Fri by appointment only, 9am-5pm. Teaching times 10am-2.30pm.

Entry Suggested donation £2 per child. Building tours £4 per person.

Yearly event or point of interest Sustainability and green living open day in September.

Directions From the A34 Milton interchange, follow signs for Milton Park; ignore signs for Sutton Courtenay village on left. Go straight across two mini-roundabouts. At the third roundabout, turn left and the centre is on your right, through some large metal gates. **Bus:** Oxford bus company / Stagecoach to Milton Park. **By train:** Nearest station Didcot Parkway – two miles.

Description An inspiring place to learn about the world we live in through fun, hands-on, experiential activities. The Centre is open all year for pre-booked group visits from schools and adult learning groups, as well as teacher training. The stunning building makes use of environmentally sustainable products from top to bottom, and is set in a nature reserve.

YOU CAN SEE

Biomass boiler • Biofuel • Thermal mass • Photovoltaics • Solar heating • Passive solar design • Mud-block building • Timber-framed building • Renewable insulation • Compost toilet • Rainwater collection • Greywater system.

SUFFOLK

81. Assington Mill

Assington Mill, Assington, Suffolk, CO10 5LZ.
Tel: 01787 229955.
info@assingtonmill.com www.assingtonmill.com
Open Open to the general public by appointment.
Accommodation B&B.

Yearly event or point of interest Assington Mill is situated in one of Suffolk's hidden valleys, out of sight of any other buildings, down a dirt track half a mile from the nearest tarmac road. The buildings are in the centre of an 86-acre farm which is managed as a private nature reserve.

Directions **Train**: Colchester station (fairly frequent). Then take a taxi. Pick up can be arranged with advanced notice. Alternatively, change at Marks Tey and take the branch line to Sudbury, stopping at Bures. No taxis but pick up can be arranged with advanced notice. **Car**: See website for directions.

Description Assington Mill runs short craft courses, lasting from one to five days, on an eclectic range of over forty subjects. Groups are typically about twelve students to a tutor, but are often much smaller. See full details on the website. Fees include home-made lunch, biscuits and cakes, sourced locally where possible.

YOU CAN SEE (BY APPOINTMENT)

Straw-bale building • Modern earth building • Timber-framed building • Renewable insulation • Compost toilet.

COURSES

A Day with Badgers Spend an afternoon and evening studying badgers: how to tell a badger sett from a rabbit hole; what tracks they make; their habits and lifecycle. You may even see one. Limited numbers.

Beekeeping for Beginners Students will learn how to set up an apiary, what equipment to buy, beekeeping terminology, the life cycle of the bee, how to identify the different members of the hive, the bee calendar, and diseases.

Cane and Rush Seating for Chairs Bring your own furniture and learn how to renew the seating using natural materials cane, seagrass and rush, under the supervision of a professional.

Dowsing and Water Divining Discover or develop your dowsing skills and knowledge in the pursuit of water; estimate the quality, depth and volume of flow. Leader: Geoff Crockford.

Feng Shui, an introduction Feng Shui is the ancient Chinese art of placement:

from an individual object in the home or workplace to the positioning of a building, it shows us how to support the flow of energy in our lives to augment health and prosperity.

Food for Free Walk around the farm to learn about the edible roots, stems, leaves, flowers and fruits of our common plants, and cook some of them later.

Fruit Tree Pruning 1-day course covering the principles of pruning, the use of tools, sharpening, plus demonstrations on how to prune new trees and old neglected ones.

Hedgerow Basketry Learn to make a stake and strand round basket with cut out handles, using at least four different weaves and a border, from completely natural materials. In a wide range of interesting colours and textures.

Hen Keeping for Beginners This course is designed for those who want to keep a few hens in the garden, and will be a mixture of theory and practical.

Life Balance Learn to re-assess your life and even transform it. Consider what changes are needed to achieve a balance. This is a new and energising 1-day course to enable you to go forward with enthusiasm and new skills.

Spinning by Hand Learn the basics of hand spinning using raw fleece from rare breed sheep and the processes involved in preparation. All materials and wheels provided but bring your own if you have them. Limited numbers.

Straw-Bale Building Learn how to build using straw bales, under the tuition of Barbara Jones, Britain's foremost straw-bale builder.

Woodcarving, an introduction 2-day introduction to woodcarving: sharpening, chisels, safe tool control. The aim is to take home a finished piece.

Many other courses, e.g. **Furniture and Book Restoration, Gilding, Coracle Making, Digital photography, Digital image editing, Plumbing for DIYers, Lawnmower servicing**.

82. Flatford Mill Field Studies Centre

 4 WORKROOMS, LIBRARY

East Bergholt, Suffolk, CO7 6UL.

Tel: 0845 330 7368, Fax: 01206 298892.

enquiries.fm@field-studies-council.org www.field-studies-council.org

Open Not open to the general public.

Accommodation 70 beds.

Directions **Train:** 4 miles from Manningtree station, with frequent services from London Liverpool Street, the Midlands and East Anglia. Taxi service available. **Car:** The Centre is 2 miles (3.5km) from the A12, between Colchester and Ipswich.

Description See 'Field Studies Centres' in 'National Organisations' at the back of the book. Eco centre status.

YOU CAN SEE

Weather station.

COURSES

A Weekend on Badgers This course is for anyone with a strong amateur interest in the identification, natural history, ecology and conservation of badgers. It also covers badger vocalisation, the controversy regarding their connection with the spread of bovine TB, and the effects of the 1992 Protection of Badgers Act. Field excursions include an evening visit to a site where badgers are seen regularly, and a daytime visit to look at setts, tracks, signs and other evidence of their presence.

Bird Songs and Calls for Beginners: the Early Spring Weekend Develop your knowledge of a variety of different bird songs and confidence in being able to recognise them. Listening in the field and reviewing sounds using DVDs will help you to memorise those elusive sound patterns.

Dormouse Ecology and Conservation A course for anyone with a professional or a strong amateur interest in the dormouse, which is fully protected under UK and European law. It is suitable for countryside professionals and managers interested in Biodiversity Action Plan species and anyone wishing to work towards an English Nature licence to handle dormice.

Environmental Art for Families No previous experience necessary! On this course we work both individually and collaboratively in a variety of fun ways, creating artwork with natural and found materials. Studio activities will include making our own paint and drawing materials, and looking at sources of inspiration.

Garden BirdWatch: Identifying Garden Birds by Sight and Sound This course is designed especially for anyone involved in the Garden BirdWatch project (organised jointly by British Trust for Ornithology and CJ WildBird Foods) or just generally interested in garden birds. By learning about bird characteristics and behaviour and seeing birds close up, you'll gain the confidence to identify both the common and scarcer species, ensuring that any data submitted to BTO is as accurate as possible.

Grass Identification for Beginners Many people shy away from naming grasses as they perceive them to be difficult, but once their structure is understood, the similarities and differences between the various species become clear.

Hedgerow Basketry This weekend introduces the traditional craft of basket weaving using some of the fascinating materials found in wild places and gardens. Explore the local woodlands and hedgerows to identify and gather suitable cuttings, and learn to build your own traditional frame basket.

Identifying Beetles This course is designed for amateur naturalists, teachers and countryside professionals interested in learning more about this amazingly diverse group of insects. A combination of field excursions, workshop sessions and illustrated talks will provide experience of the methods of collecting, identifying and preserving beetles.

Identifying Bumblebees This short course is designed for amateur naturalists, teachers and countryside professionals. It is particularly suitable for anyone involved in biological recording schemes at a local or county level. Through field excursions you become familiar with the key identification features of at least six of the most common species of bumblebee, as well as two or three species of parasitic cuckoo bumblebee.

Identifying Dragonflies and Damselflies This short course introduces identification, life histories, behaviour and conservation using field excursions complemented by indoor activities, and a comprehensive collection of photographs and specimens. Designed for amateur naturalists, teachers and countryside professionals, and anyone involved in biological recording schemes at a local or county level.

Identifying Moths and Butterflies A weekend introducing the identification and natural history of this insect group. Only live viewing techniques are used: light traps and sugaring for moths, and sweep nets for butterflies. Working through the moth catch in the morning is complemented by afternoon excursions to look for butterflies.

Identifying Spiders Learn about the key identification features, natural history and ecology of the spider. Field excursions to local habitats will introduce a range of techniques for finding and collecting spiders.

Mammal Identification Learn to identify Britain's land mammals from sightings and field signs (including calls, droppings, feeding remains and footprints). With lectures, slide shows and evening activities, as well as hands-on experience of doing owl pellet analysis and Longworth live-trapping for small mammals.

Microscopy for Beekeepers This practical weekend course explores a number of aspects of beekeeping using low- and high-power microscope techniques. These include looking at flower morphology in relation to pollination, preparing microscope slides of honey bee exoskeletons and of pollen from flowers, pollen loads and honey. There will be practice in dissecting the honey bee and looking for evidence of specific diseases and parasites.

Practical Hedgelaying Skills Learn how to lay hedges using traditional tools and materials. Each autumn in recent years we have worked on a number of hedgerows around Flatford Mill, so you will be able to see how a laid hedge develops and improves with time. Illustrated evening talks complement the practical activities, and safe working practices are emphasised throughout.

Simple Baskets for Gathering and Picking Make your own beautiful rustic willow baskets. Possible projects include a flower trug, drying tray, berrying basket, apple basket or fruit dish. All materials provided.

The Rustic Gardener: Introducing Practical Woodcraft This course explores how you can grow and harvest your own materials from woods, hedgerows, gardens or allotments, and develop practical woodworking skills to create a range of useful tools and furniture.

Trees and Tree Identification: the Challenge of Early Winter In late autumn and early winter foliage and fruits of broad-leaved trees are a blaze of colour. By the end of November we have only twig formation and silhouettes to help us identify these trees. This course shows you how.

'Trugs to Trellis': Willow Craft for Gardeners Learn to make simple items out of willow such as trugs, wigwam plant supports, hanging baskets, decorative trellis or garden sculptures. Expect to complete at least two projects and leave with plenty of ideas for further works. No previous experience necessary.

A wide range of art courses is available in this 'Constable Country' Field Studies Centre.

Further courses on website – e.g. **Woodland History**, more **Species Identification, Plant Identification.**

83. Westley Bottom

Bury St Edmunds, Suffolk, IP33 3WD.

Tel: 01284 747500, Fax: 01284 747506.

www.nationaltrust.org

Open Bookings only.

Directions Will be supplied at time of booking.

Description National Trust Head Office.

See 'National Trust' in 'National Organisations' for more information.

YOU CAN SEE (BY APPOINTMENT)

Wood energy scheme.

SURREY

84. BioRegional Development Group

    OD MON-FRI

BedZED Centre, 24 Helios Road, Wallington, Surrey, SM6 7BZ.

Tel: 020 8404 4880, Fax: 020 8404 4893.

info@bioregional.com www.bioregional.com

Open Mon-Fri 9.30am-5pm. Please pre-book CPD training and tours (charges apply).

Entry Free.

Yearly event or point of interest We take part in Open House each year. Please visit the Open House website (www.openhouse.org.uk) for details.

Directions **Train:** From London Victoria, alight at Hackbridge, turn right out of station. We are 5 minutes along on the right hand side. **Tram**: Alight at Mitcham Junction, turn left out of the station, and we are 10 minutes along on the left hand side. **Bus**: Take the 127 bus from Tooting or Purley, which stops right outside. We prefer visitors to travel by public transport, but if that's not possible, park at Hackbridge rail station and walk as shown above.

Description BioRegional is an independent environmental organisation dedicated to developing practical solutions for sustainable living. It runs an exhibition about sustainability, showcasing projects which range from paper and textiles to forestry and sustainable communities like BedZED. Visitors can also explore the BedZED eco-chic show home, join guided site tours and attend continuing professional development training sessions.

YOU CAN SEE

Biofuel • Thermal mass • Specialist ventilation • Photovoltaics • Passive solar design • Rainwater collection • Greywater system • Brick and thermally massive concrete block.

COURSES

One Planet Living Training for Construction Industry Professionals

1-day courses on sustainable development. There are three seminars in the morning, followed by a workshop in the afternoon using the 10 One Planet Living Principles (www.oneplanetliving.org) as a framework to design a sustainable development. The course counts for 7 hours of CPD. For more info and a booking form, see the website.

85. Juniper Hall Field Studies Centre

   5 WORKROOMS, LIBRARY

Dorking, Surrey, RH5 6DA.
Tel: 0845 458 3507, Fax: 0845 458 9219.
enquiries.jh@field-studies-council.org www.field-studies-council.org

Open Not open to the general public.

Accommodation 80 beds.

Directions **Train:** From London (Victoria and Waterloo) and Horsham, all trains stop at Dorking. Taxis are readily available outside the main station. Some trains stop at Boxhill and Westhumble station, although taxis are unavailable here. The centre is only a short walk away (about 1 mile). Trains from Reading and Guildford or Gatwick, Redhill and Reigate stop at Dorking Deepdene station, which is a short walk from the main station. **Bus:** The 465 service runs from opposite Dorking station and stops just outside the centre (Headley Lane stop), south of the Mickleham Church stop. The route runs from Teddington and Kingston to Dorking via Leatherhead. **Car:** See information on the website.

Description See 'Field Studies Council' in 'National Organisations' section at back of book. Eco Centre status.

YOU CAN SEE

Weather station • Conservation and teaching garden • Butterfly garden • Invertebrate study area • Woodland met station (small).

COURSES

Bats and Bat Surveys: A Foundation Course for Environmental Consultants Changes in environmental legislation, survey licensing arrangements and PPS9 have all increased the demand for bat surveys and the need for consultants with knowledge about bat survey work. To ensure this demand can be met effectively, the Bat Conservation Trust is conducting these highly developed courses specifically for environmental consultants. The courses include tuition on: bat biology and ecology; legislation and policy; surveys methods; assessing value and impact; mitigation and monitoring, and recognising when to recruit a more experienced bat worker.

Bird Watching for Absolute Beginners This weekend is primarily for those who are about to start or have just started birdwatching. It will cover topics that are of greatest interest to those that have yet to birdwatch in the field. Indoor sessions will include the use of slides, video and birdwatching equipment.

Chasing the Purple Emperor At the peak time for woodland butterflies, we will introduce you to their ecology and investigate the local countryside and varied habitats around Juniper Hall. Moth trapping will enable us to compare and examine how moths differ from butterflies. We will explore butterfly life cycles, adult and larval behaviour, mating strategies, distribution and major habitat management requirements.

Family Bushcraft Weekend Teaches some of the fundamental skills needed to survive in the outdoors. Learn to: use bushcraft tools; make some useful items; collect and purify water; the principles of fire lighting; how to put up a hammock (including all the necessary knots). Build a shelter from materials found in the woods and learn good environmental practice. Finally the focus shifts to the natural world and the plants and animal signs around us.

Grass Identification in Spring This course is designed to enable anyone to identify grasses by their vegetative (non-flowering) characteristics, and should be especially valuable to teachers of ecology and those engaged in botanical survey work, as well as of interest to amateur botanists.

Introduction to the Flowers of the Chalk Suitable for beginners, this short course offers an introduction to one of Britain's most species-rich habitats. Juniper Hall lies in the midst of the Box Hill Site of Special Scientific Interest, also designated as a European Special Area for Conservation, and has several excellent chalk grassland sites close by, managed for maximum biodiversity by the National Trust.

Lichens near London This course has been planned as an introduction to the lichens of semi-polluted regions and will cover their identification, distribution and growth. Other aspects will include the uses of lichens especially for monitoring air pollution, and photography.

Spring Birds and their Songs The Surrey countryside is alive with bird song in spring. We will visit many local sites, listening to and learning to recognise songs and calls. In addition to a dawn chorus walk, we will visit Bookham Common for its nightingales and Thursley Common for several specialities, including Dartford warbler. There will be opportunities to sound record the birds.

Tatting Weekend Tatting is an old, simple and portable thread craft producing a strong lace-like fabric. This weekend course is suitable for beginners as well as more experienced students and is designed to inspire you to create interesting butterfly motifs, 3D flowers and personalised greetings cards.

Wild Flowers for Beginners This short course covers the structure of plants and flowers, concentrating on those features used in identification. Field excursions will explore the species-rich chalk grassland and deciduous woodlands which surround Juniper Hall. Informal evening sessions will include illustrated lectures and a chance to look at some flowers in more detail using binocular microscopes.

Many other courses e.g. **Dragonflies, Orchids, Small Mammals** – see website.

CHILDREN'S COURSES

Minibeasts, Mice and the Mole A chance to take a closer look at the animals around us which we rarely see. How high can a mouse jump? Why does a vole have to be careful with its toilet habits? How do freshwater invertebrates remain in freshwater? The course uses an investigative approach, and concentrates on small mammal trapping in a variety of habitats, including grassland and woodland; invertebrate hunting in grass, woodlands and leaf litter, and river dipping for invertebrates.

Also **Animals All Around Us**

86. Surrey Wildlife Trust

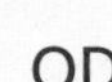
OD

School Lane, Pirbright, Woking, Surrey, GU24 0JN.

Tel: 01483 795440, Fax: 01483 486505.

info@surreywt.org.uk www.surreywildlifetrust.org

Open 7 days a week 9am-5pm.

Entry Free (donations welcome). Some walks, talks and events do need pre-booking, but most do not.

Yearly event or point of interest Members meeting with AGM. Going Wild at Bay Pond (family event with lots of activities). Bat walks throughout the summer and autumn.

Description Surrey Wildlife Trust is a registered charity working to conserve the countryside and wildlife of the county. With over 60 sites across the county, it currently manages 11,500 acres of Surrey's countryside and is one of 47 partners in the Wildlife Trusts, the largest organisation working exclusively in the UK on all aspects of nature conservation.

YOU CAN DO

Walks and tours • Family activities – see website.

COURSES

Refer to the website for good range of weekly events for training and walks / talks where you can learn about biodiversity, wildlife and woodlands.

EAST SUSSEX

87. Brighton Peace and Environment Centre

    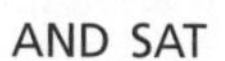 OD TUES, WEDS, THURS AND SAT

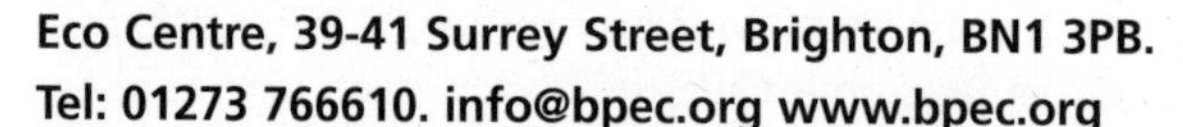

Eco Centre, 39-41 Surrey Street, Brighton, BN1 3PB.
Tel: 01273 766610. info@bpec.org www.bpec.org

Open Tues, Weds, Thurs and Sat, 10am-6pm.

Entry Free. Become a member to have 'full' access. Please be aware that some events / tours need pre-booking.

Yearly event or point of interest Extensive library providing information on the issues we promote (as above).

Directions The centre is directly opposite Brighton train station.

Description As a Development Education Centre, BPEC promotes peace, social justice, sustainable development and environmental conservation through education. BPEC responds to the growing interest within Sussex by providing information and education for the community and local schools and also works with community groups, local businesses and the public. Promoting campaigns on global issues, it is an organisation run by volunteers.

Eco market Not held at the Centre, but venue and info are on the website: e.g. organic and fair trade gifts, advice on environmental topics, fresh organic farmers produce, solar panels and reclaimed wooden furniture, eco non-toxic paint, local artists, healthy concerts in heated dome, café, with emphasis on local producers to help minimise transportation pollution.

YOU CAN DO

Volunteering • Environment and Sustainability Work Group.

COURSES / EVENTS

Eco-Craft Classes Recycle your way to stunning jewellery, gifts, cards and decorations.

Global Citizens Advocates A 3-day course for teachers to inspire them to become advocates of Global Citizenship, so that the principles and ideas can be embedded in the school environment. The course is run three times a year, and is open to teachers and educators from Brighton and Hove, and East and West Sussex. Participants will also receive some free teaching resources.

Monthly discussion group Topics listed on the website.

Introduction to Permaculture

88. Low Carbon Network

OD MON-FRI

Earthship Brighton, Stanmer Park Nurseries, Stanmer Park, Brighton, BN1 9PZ.
Tel: 0845 680 0015. info@lowcarbon.co.uk www.lowcarbon.co.uk

Open Mon-Fri 10am-4pm. Regular tours on 1st and 3rd Sunday of every month.

Entry £10 per person per technical tour. £5 for regular tour (please pre-book).

Yearly event or point of interest Earthship Brighton – visit the only Earthship in England. Earthships are 'green' buildings, constructed using waste car tyres, which use the planet's natural systems to provide utilities: using the sun's energy and rain to provide heat, power and water. Earthships give a glimpse of a luxury, low-carbon future, and how a zero-waste and zero-energy society might look.

Directions Car: From Brighton city centre. Follow A27 (Lewes Road) out of town. Take Stanmer Park turning just before the University of Sussex. Drive through park past Stanmer House. Take a left at the 'T' junction. Straight on through the double gates and follow the road to Stanmer Organics. Contact for information on public services.

Description The Low Carbon Network is a not-for-profit company formed in 2001 by people living in Brighton to raise awareness of the links between buildings, the working and living patterns by people that use them, and climate change. The Network aims to inspire cultural changes in the design of more conventional buildings through extensive communications work and highly innovative building projects that demonstrate best practice in ecological design and environmental construction.

YOU CAN SEE

Biomass boiler • Thermal mass • Specialist ventilation • Photovoltaics • Solar heating • Passive solar design • Wind turbine – domestic • Earthship • Compost toilet • Rainwater collection • Greywater system • Reed bed.

YOU CAN DO

Volunteering.

89. Weald Woodnet

Woodland Enterprise Centre, Hastings Road,
Flimwell, Wadhurst, East Sussex, TN5 7PR.
Tel: 01580 879552, Fax: 01580 879552.
info@woodnet.org.uk www.woodnet.org.uk

Open to the general public by arrangement.

Yearly event or point of interest Open days, exhibitions often associated with Architecture Week etc. *WoodLots* magazine for timber and wood services, produced every two months: www.woodlots.org.uk.

Directions On A21 between Hastings and Tunbridge Wells. Entrance ½ mile south of Flimwell crossroads.

Description Weald Woodnet provides a network in south-east England to link growers of wood with users of wood, and to support use of local timber in sustainable construction projects, renewable energy and for landscape / furniture designers. Woodnet provide advice on sourcing sustainable local timber, and run training courses on woodland management and home-grown timber use at Flimwell.

YOU CAN SEE (BY APPOINTMENT)

Biomass production – woodchip • Biomass boiler • Passive solar design • Timber-framed building – chestnut gridshell • Renewable insulation.

COURSES

Examples of subject areas are:

Art – painting, sculpture and photography • Coppicing and Traditional Woodland Skills • Health and Safety in Woodlands • Local Timber in Construction and Architecture • Local Wood for Heat: a secure and carbon neutral fuel source • Local Timber in Design: a versatile renewable material • Locking up Carbon: the role of trees and timber • Mobile Sawmilling and Wood Chipping • Silviculture and Measuring Trees • Understanding the Working Woodland Environment • Woodland Archaeology and History.

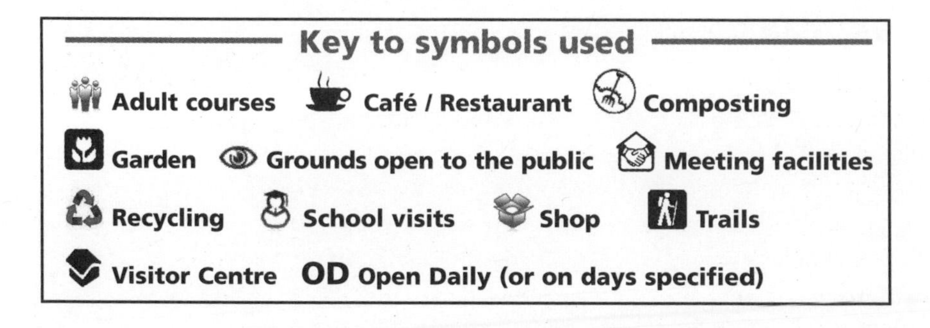

WEST SUSSEX

90. Future Roots

Stanmer Park, just off the A27 between Brighton and Lewes.

Mobile Tel: Matt Boysons 07799 630027, Mark Whittaker 07776 101266.

Mattboysons@yahoo.co.uk mark@futureroots.co.uk www.futureroots.co.uk

Open by appointment only.

Directions Available by contacting us by email or phone.

Description At Future Roots it is believed that a sustainable way of life can be achieved using low-impact methods, and that the perfect structure for this way of life is the yurt. Working with wood dictates a certain pace to life; a tree grows slowly, is selected and cut carefully, and finally becomes something new and worthwhile only because time has been taken over its manufacture. Future Roots have embraced this gentle, proper pace and feel that their yurts carry with them that atmosphere, whether you are buying or renting one. It is not without good reason that there are still millions of people living in yurts and gers across Asia in the harshest and most extreme of weather conditions. As shelter from those elements and as the perfect nomadic home, they work incredibly well and have done so for over a thousand years. There are even people living in them all year round across Britain and Ireland, so it's fair to say they work in our joyous weather as well.

COURSES

The Future Roots Yurt-Making Course 4 days over two weekends. Covers everything involved in the construction of a yurt. Our aim is to give students enough knowledge and self-confidence to build a yurt of their own. The course includes instruction and practice with all the tools used to produce a quality yurt frame, and how to source and select suitable timber. We work on a ratio of two instructors to up to four students. We cover many skills and techniques on the course, but it should not be beyond the ability of any reasonably able person to complete. Lunch is not supplied, but there are facilities for barbequeing. Tea, coffee and hot water are provided.

91. Small Farm Training Group

34 School Lane, Ashurst Wood, West Sussex, RH19 3QP.

Tel: 01342 825453. gto@sftg.co.uk www.sftg.co.uk

Open Not open to general public – membership organisation. Social events, monthly meetings, attendance at local agricultural shows.

Yearly event or point of interest Attend Heathfield Show (May) Smallholder's Show (July) and South of England Autumn Show (Sept-Oct).

Directions Held at various locations: please contact for advice.

Description Sussex-based group which provides inexpensive training courses for anyone with an interest in smallholding and countryside skills. Membership of group gives access to unlimited, unbiased advice and quarterly newsletters. A small charge is made for each course, which come from member's suggestions and are designed for both beginners and more experienced.

COURSES

Courses are run in the evening or at the weekends throughout the year (except August). The wide range of courses includes:

Countryside skills Charcoal making • Coppicing • Hedge-laying • Productive woodland • Stock fencing • Introduction to smallholding • Baler maintenance • Chainsaw maintenance.

Machinery Tractor maintenance • Trailer handling.

Gardening Allotments for beginners • Fruit growing • Fungi identification • Herb gardens • Propagation • Top fruit pruning • Vegetable growing.

Livestock Beekeeping • Cattle • Goats • Introduction to sheepdogs • Pigs • Poultry.

Sheep Sheep shearing.

Other Cheese making • Sausage making.

92. South East Wood Fuels Ltd

South East Wood Fuels Ltd, Premier House,
Shoreham Airport, West Sussex, BN43 5FF.
Tel: 01273 440943. info@sewf.co.uk www.sewf.co.uk

Directions Courses at various locations – see website for details.

Description South East Wood Fuels Ltd is a not-for-profit company working to supply reliable, standards-based wood fuel in the South-East. It is developing a broad network of suppliers from across the South-East that are trained in the production and supply of quality wood fuels. Through this it offers unrivalled levels of supply, quality and security.

COURSES

South East Wood Fuels offers training in seminar format in the standards, quality assessment and measurement of woodchip for the wood fuel industry. The days include an overview of current technology (chippers, boilers, etc.); the processes involved in producing fuel to a standard suitable for modern boilers; monitoring and testing; and the practicalities of storage, handling and delivery. This training is aimed at existing forestry contractors and woodland owners looking to produce quality wood fuels for this growing market. The day is also of value to all those involved in the planning, installation and management of woodchip heating systems.

On completing this course, attendees will: understand the European woodchip fuel standards; understand these standards in relation to woodchip boiler technology and operation; understand methods for the measurement and monitoring of woodchip supplies, including the use of moisture meters; understand production processes and the infrastructure necessary to produce and deliver woodchip to a specified standard; have a training pack to act as a reference; and a certificate of attendance. See website for dates.

93. Weald and Downland Open Air Museum

    OD VARIES THROUGH SEASONS

Weald and Downland Open Air Museum, Singleton, Chichester, West Sussex, PO18 0EU. Tel: 01243 811363, Fax: 01243 811475.

office@wealddown.co.uk www.wealddown.co.uk

Open to the general public. Opening times vary: see website.

Yearly event or point of interest Sustainability Event – Learn more about a wide range of low impact, 'green' products and techniques for building and sustainable living in the 21st century. Includes seminars, demonstrations and a wide variety of exhibits to interest domestic users and professionals. Other special events: Tree Dressing Day, Heavy Horse Day, Autumn Countryside Show, and more.

Directions The Museum is situated 7 miles north of Chichester on the A286. **Bus:** The museum is served by Stagecoach Coastline buses on the number 60 route between Bognor Regis, Chichester, Midhurst, Haslemere and Guildford. **Train:** The nearest rail station is Chichester.

Description Set in 50 acres of Sussex countryside, this is a fascinating collection of nearly 50 historic buildings dating from the 13th to the 19th century, many with period gardens, together with farm animals, woodland walks and a picturesque lake. Rescued from destruction, the buildings have been carefully dismantled, conserved and rebuilt to their original form and bring to life the homes, farmsteads and rural industries of the last 500 years.

COURSES

A Dowsing Workshop Explore the background and practise the ancient craft of dowsing.

An introduction to charcoal burning A practical day exploring charcoal burning using both traditional and current methods, from selection of wood to lighting the kiln, to marketing the product.

An Introduction To Dating Timber-Framed Buildings Roof timbers often provide the best evidence for the period of construction. We use the Museum buildings to look at roofs dating from the 1400s to the 1900s.

Beekeepers' Preparation for Winter Find out when and how to take off honey stores in autumn, reducing the structure of the hive and feeding to prepare the bee colony for the coming winter. A useful follow-on from Beekeeping for Beginners (below).

Beekeeping for Beginners An introduction to keeping bees. Investigate the components of a beehive, see how to make up beeswax frames, and find out what equipment you will need.

Birds of Prey Experience For those who are considering taking up falconry or hawking, or would simply like to enjoy a day out that offers something completely different.

Charcoal Burning A practical day exploring charcoal burning using both traditional and current methods, from selection of wood to lighting the kiln and marketing the product.

Cob Walling – History, Theory and Practice The day school will explore the various types and methods of cob wall construction in the region. It will also examine causes of failure, repair strategies and problems relating to alterations to cob structures. Includes some hands-on practice.

Construction and Repair of Timber-Framed Buildings Exploration of the background to to the construction of timber-framed buildings and the dating of buildings by their construction techniques. The course will also explore structural problems and sympathetic remedial methods.

Continuous Hurdle Fencing Learn the practicalities of making a continuous wattle fence using mature hazel rods woven between upright supports. Advice on sources of materials, tools needed etc.

Coracle Making Weekend Coracles are traditional riverboats. Round in shape, they are paddled with one oar. On this 2-day course you will make a traditional ash slatted coracle to take home, and try out your boating skills on the Museum's millpond.

Corn Dolly Workshop Learn the history and development of this ancient craft, as well as the practical skills involved in weaving corn dollies.

The Cottage Herb Garden – for Beginners Find out how the cottage herb garden evolved to supply the needs of the family, and how this can be interpreted in your own garden today. The day includes planning and maintaining a herb garden, integration with flowers or vegetables, techniques for successful herb growing, and an introduction to the use of herbs in cookery, fragrance and home medicine.

The Cottage Garden – Herbs for Health The origins of the herbaceous border lie in growing herbs for home remedies in the past. Led by a medical herbalist, the course offers guidance on the most safe and useful herbs to grow and how to harvest and use or preserve them. A practical day, including the opportunity to make a footbath, herbal honey syrups, herbal teas, infused oil, a herb pillow and more.

Energy Conservation in Traditional Buildings The day will include looking at the implications of improving energy efficiency for traditional buildings, a review of the relevant regulations and guidance, case studies and a practical session of carrying out an air pressure test.

Garden Gate from Scratch Make your own garden gate to take home in two days. On the first day, the frame will be made using mortice and tenon joints. On day two, rails will be prepared and fixed to the frame.

Gate Hurdle Workshop Students on this workshop will make a gate hurdle traditionally used for penning sheep, using cleft ash and hand tools.

Green Architecture Reduce the environmental impact of building at design stage. A day school for architects and designers concerned with sustainability issues in new-build and refurbishment.

Greening your Garden An introduction to environmentally sensitive gardening techniques to help make your garden more productive, wildlife-friendly and enjoyable. Each day will consist of four sessions.

Greening Your Home Two separate workshops providing advice, information and inspiration for improving the sustainability of your home and taking care of the environment.

Herbs for the Kitchen An appreciation of cottage garden herbs with details of when to gather and how to dry and store them and recipes past and present. Some hands-on experience included.

Intermediate Timber Framing – Wall Framing (follow-on course from Timber Framing from Scratch) This 5-day intermediate framing course is designed for those who have already attended a 'Timber Framing from Scratch' course and who wish to learn more about the marking and cutting of studs and braces. The frame worked on will be one that has recently been made on a Timber Framing from Scratch course.

Intermediate Timber Framing – Roof Framing A 5-day practical course for students who have attended the Timber Framing from Scratch course. The common, principal, hip and jack rafters are marked, cut and fitted to a timber frame that was made on a previous course. All completed roof members are pitched on the last afternoon.

Irons in the Fire A practical half day in the Museum's 150 year old working smithy, learning about the traditional skills of the village blacksmith. Each student will make their own simple object to take home.

Make a Bat Box Why not encourage these fabulous crepuscular creatures into your garden by providing a box for them to roost in, and brush up on your woodworking skills at the same time?

Make a Birdbox Give the birds in your garden a new home. Using hand and power tools, you will make one or more boxes suitable for a variety of garden birds.

Make a Bentwood Chair Learn how to make a comfortable, natural bentwood chair to take home, selecting your own new timber and using a minimum of tools to facilitate the steaming process.

Oak Shingles: History, Manufacture and Use A day of talks, demonstrations and practical experience of the equipment used in cleaving and finishing oak shingles under the guidance of the most experienced steeplejacks in the country.

Secrets of the Tudor Stillroom The stillroom was the source of a variety of potions, salves, pest repellents, cleansers, cosmetics and fragrant mixes, in addition to the all-important distilled aromatic waters. Explore some of the recipes handed down over the generations and make cough sweets, salves, herb honeys and drinks.

Skep Making Learn to make a useful addition to the beekeeper's equipment from straw, and find out how skeps are used in beekeeping. A rustic feature for a cottage garden, even if it is not inhabited by bees!

Small Farm Animals – Sheep and Pigs A day school to explore the feasibility and management of a smallholding with sheep and pigs, including feeding, housing, breeding, and regulations.

Soap and Soap Making Find out how people lived without modern detergents. A day-school on Tudor hygiene and cleanliness, learning how to make lye and soap, and the uses of soapwort.

The Singleton Shingled Spire Thousands of steep spires were added atop medieval church bell towers, replacing the simpler pitched roof. Over three-week courses we frame and erect on the Museum site a 14-foot-high spire based on a local example. The resulting polygonal frame is used as the model for teaching the craft of shingling.

Timber Identification of Species Introduction to the identification of timber species through examination of anatomical features, demonstrations and practical work using hand lenses and microscopes.

Timber Framing from Scratch A 5-day practical course introducing students to the historic use of structural oak framing, tools and techniques. The posts, cills, plates and beams of a ten foot square timber frame are prefabricated during the course using only traditional tools and techniques, and the frame is erected on the last afternoon.

Traditional Lime Plasters and Renders A practical 2-day course covering the fundamentals of lime plastering, from the simplest renders to the finest ornamental work. Lectures followed by practical demonstrations, hands-on experience and opportunity for discussion.

Traditional Ropework Come and try your hand at the art of traditional rope work. Students will make a round clout mat and a sailor's whisk brush during the day.

Wattle and Daub Insights into the historic use of wattle and daub, and its repair and conservation today. A morning of touring the Museum's examples followed by a practical exercise applying wattles and daubing them.

Wildlife and the Law A day for anyone who needs to understand the legislation relating to wild creatures and plants, including the Wildlife and Countryside Act 1981 and the Crow Act 2000.

Willow for the Garden Make a living willow structure to plant in your garden.

Willow Workshop Using English brown willow and traditional techniques, learn to 'weave and wale' a basket to take home.

CHILDREN'S ACTIVITIES

Activities during half term and the summer holidays.

94. Willow Crafts with Ganesh Bruce Kings, Elaine Kings and Geoff King

Plovers Cottage, Batchmere Lane, Almodington, Nr Chichester, West Sussex, PO20 7LJ. Tel: 01243 511052, Mobile: 07779 566951.

ganesh_willow@hotmail.com www.creativewillow.com

Directions Courses held at various locations – see website for details.

Description Willow is used as a means to teach life skills to people of all ages and backgrounds. The course leaders take living willow, organic growing techniques and other projects into schools, and teach a variety of courses from their own garden, so people can learn new skills alongside building confidence and self-esteem. They also supply both living and non-living bundles of willow. "Most of our workshops are at our home either in our studio or our garden. Prices include tuition, a full organic vegetarian lunch and all hot and cold drinks. Come and enjoy a day learning a new skill and sharing good company and good food."

COURSES

Festive Craft and Christmas Tree Workshops Make an alternative Christmas Tree Teepee this year, and use it again in the spring for your sweet peas! Plus Christmas stars, wreaths, hearts, and lots of different ideas for Christmas presents.

Living Willow Seat Workshop Make a living willow seat/chair to plant in the garden using our home-grown living willow uprights, and cut willow rods to weave the base, sides and back.

Willow Craft Workshop Make a vase, heart, bird-feeder, hanging plant holder, dragonfly, stick platter, shooting star, snail, or snake – the list is endless.

Willow Basket-Making Workshop A chance to make a small round or oval basket using traditional techniques from a variety of English willow.

Willow Teepee, Hurdle and Trellis Workshop Make a traditional tepee, hurdle, trellis or plant supports for your climbing plants, using different varieties of willow.

FOR SCHOOLS

Living Willow Projects Students participate in making a living willow structure for the school. During the first growing season there will be further visits to the school for willow maintenance, staff training and interactive storytelling with the children. Further maintenance can be negotiated.

Apple Pressing Day Hands-on interactive apple pressing and preservation of juice.

Driftwood Sculpture Workshop Making sculptures out of collected driftwood, shells and seaweed.

Environmental Grounds Day Making bug hotels, bird feeders and bird boxes to help the wildlife in our gardens.

Hand-Dyed Rainbow Candle Workshop How to colour-dye a white candle rainbow, in the correct sequence.

Interactive Storytelling Ecological storytelling for children.

Non-living Willow Sculpture This can be as abstract as the school desires: works include dragons, dragonflies, a scarecrow, an earth goddess, a wild wizard, a school guardian, a pond guardian.

Organic Veggie Growing in Schools Making practical raised beds and filling with good quality organic matter and making compost bins.

Storytelling Benches Making seats and benches out of recycled oak fence posts with older children using tools.

Willow craft for schools Children make a 2D willow shooting star, 3D bird feeder, snail or vase (with plastic test tube). We aim to work with a maximum of three full classes, or two classes in much more detail.

SOUTH-WEST ENGLAND

CORNWALL

95. Botallack Count House

Trewellard, Pendeen, Nr St Just, Cornwall, TR19 7SX. Tel: 01736 786156.
www.cornwall-calling.co.uk/national-trust/botallack-count-house.htm

Directions near Levant Mine and Beam Engine. Please phone for directions.

Description See 'National Trust' in 'National Organisations' in back of the book.

YOU CAN SEE (BY APPOINTMENT)

EarthEnergy system.

96. Hugh Cole – Rustic Roundwood Furniture

2 Riverside Cottage, Tresillian, Truro, Cornwall, TR2 4BA.
Tel: 01872 520109 or mobile 07764 156404.
rusticfurniture@hughcole.co.uk www.hughcole.co.uk

Directions Please contact for directions.

Description Traditional rustic furniture has a beauty and form that is unique. The use of hazel only adds to the beauty of the finished product. The design of these chairs makes them strong and practical, for indoor or outdoor use, and there are sizes to suit adults and children. They also produce a range of other practical items, ranging from planters to bird tables and coffee tables, and more – see website.

COURSES

Hazel Chair Course (2 days) Collect hazel from local sources and learn how to safely construct your own version of a straight-back chair. This portion of the course is much the same as detailed below for the 4-day course. Whatever you make, you take home!

Hazel Chair Course (4 days)

Day 1: Cut hazel in a local Cornish Wildlife Trust woodland on the Camel Trail and learn basic practical coppicing techniques, a little about its history and the traditional uses of the wood, plus the environmental and conservation issues. Pick the brains of Phil Harris, who is an expert on flora and fauna and a butterfly specialist.

Days 2-4: Held at the Wadebridge campus, guided by Hugh Cole. The course is spread over four days to give the chance to be involved in every stage of construction, rather than just fitting together pre-cut pieces of wood. If time permits, have a go at other woodland craft skills such as using a draw horse, rounding planes and a pole lathe. For anyone over 16, students from 17 to 73 have attended and 65% have been women. Everyone takes home a unique chair.

97. Cornwall Sustainable Buildings Trust

Currently at Eden's Watering Lane Nursery, Lobb's Shop, St Austell, PL26 6BE. Tel: 01726 68654. Fax: 01726 67028.

info@csb.org.uk www.csbt.org.uk

Directions Moving locations, so please contact for latest directions.

Description CSBT is a registered charity aiming to achieve a state where all parties are working towards the common aim of sustainability. Currently they have over 200 associates from both public and private sectors, and many interested individuals. They deliver training in a range of traditional and sustainable building skills and regularly hold interesting events. CSBT is also working closely with construction professionals, academia and the renewable energy sector to provide objective guidance to planners and developers on sustainable construction options.

YOU CAN SEE (BY APPOINTMENT)

Woodchip production • Pellet production • Biofuel • Thermal mass • Specialist ventilation • Exchange heat pump • Photovoltaics • Solar heating • Passive solar design • Wind turbine – domestic and medium • Straw-bale – possibly • Cob • Timber-framed building • Renewable insulation • Compost toilet – possibly • Rainwater collection.

COURSES

Build Your Own Windmill • Choosing Sustainable Building Material • Cornish Hedging • DIY Solar Water Heating • Hazel Hurdle-Making • Introduction to Airtight Passive Ventilation • Introduction to Renewable Energy Systems • Introduction to Bankshores • Introduction to Cob and Rammed Earth • Introduction to Cob Building • Introduction to Composting Toilets • Introduction to Photovoltaics • Introduction to Roofing • Introduction to Scantle Slating • Introduction to Wood Shingles • Lime and Other Products • Natural Garden Sculpture • Principles of Sustainable Construction • Recycled Materials and their Uses in Buildings and Homes • Site Waste Management • Straw-Bale Building Theory and Practice • Sustainable Building Materials • Sustainable Interior Design • Decorative Effects • Sustainable Marketing and Packaging • Sustainable Transport • The Practical Use of Lime • The Theory Of Lime • Timber-Frame Raising.

98. Eden Project

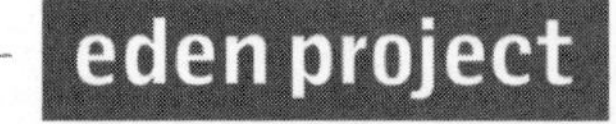

OD

Bodelva, St Austell, PL224 2SG. Tel: 01726 811911, Fax: 01726 811912.
webmail@edenproject.com www.edenproject.com

Open 7 days a week, 10am-6pm (winter 4.30pm close).

Entry Adults £13.80, Children £5, Seniors £10, Students £7, Family £34 (this gives you access to everything).

Directions **Bus/train:** St Austell train station is connected to the Centre by regular buses. **Car**: Eden is easily accessible from both the A30 and A38. Take the M5 southbound towards Exeter, then the A30. Continue on this road to the Innis Downs roundabout, from which Eden is signposted.

Description Eden Project Limited is owned by the Eden Trust, which is a registered UK Charity (No. 1093070). Eden is home to the two biggest greenhouses in the world – the Humid Tropics Biome and Warm Temperate Biome. More than just a green theme park, Eden is about connecting plants, people and places. We are a living demonstration of regeneration, and we aim to reconnect people with their environments locally and globally. We don't have all the answers, we don't want to tell others what to think; what we do is invite people to explore their world afresh.

YOU CAN SEE

Biomass boiler • Biofuel – Hydraulic fuel • Thermal Mass • Specialist ventilation • Photovoltaics • Passive solar design • Straw-bale building • Earth – cob • Timber-framed building • Renewable insulation • Rainwater collection • Greywater system • SUDS • Sustainable design is fundamental to many buildings, especially Core, which has a Glulam double curvature roof • Sustainable materials: an innovative ground tube and windsock ventilation.

Feeding the swans at Welney Wetland Centre .

Caldecotte Lake and windmill at The Parks Trust.

Bronze casting at The Yarner Trust.

Forest crafts at The Upcott Project.

Students at Musgrove Willows.

Part of the forest garden at the Agroforestry Research Trust.

A floral feast at The Organic Centre.

Straw-bale building at Carymoor Environmental Centre.

Honey harvest at The Organic College.

The Wildlife Trust's Attenborough Centre.

Mike Wye's lime-rendered cob barn.

Coppiced chestnut gridshell at Weald Woodnet.

The Wildlife Trust's Centre for Wildlife Gardening.

Summer house at Abey Smallcombe.

Roundhouse at Carymoor Environmental Centre.

The Ecohome at the CREATE Centre in Bristol.

Wind turbine at the EcoHouse.

Eco roof at The Hockerton Housing project.

YOU CAN DO

Volunteering • Eden sets great store by live performance, and where possible uses people, not technology, to engage the public. As a result there are a large number of guides, teachers, stewards, and storytellers, whose task it is to make a visit to Eden anything but a passive experience.

CHILDREN'S ACTIVITIES

There are play areas designed to make children think about their lives and how plants play an important part. There are also trails that inspire sustainable thinking – teaching about the origins of products and objects that we use each day.

99. Falmouth Green Centre

Union Road, Falmouth, Cornwall, TR11 4JW.

Tel: 01326 377173. info@fgc.org.uk www.fgc.org.uk

Yearly event or point of interest Open day once a year – see website for details.

Directions From Falmouth, follow signs for Falmouth Hospital. Turn into Kimberly Park, bear right into Trescobeas Road. At mini-roundabout turn right on to Union Corner. At mini-roundabout continue forward on to Union Road (signposted Penryn, Truro).

Description FGC is a community enterprise working to promote sustainability, in partnership with the practical conservation charity BTCV. FGC raises environmental awareness, encouraging community participation. This includes a waste-wood project, using the timber for wildlife habitat boxes and garden furniture, and a nursery project growing and selling organically grown herbs, wild flowers and native trees. The grounds have a community garden with outside classroom, wildlife and woodland area, organic plots and orchard.

YOU CAN SEE

A garden shed made out of cob • An outdoor classroom • Community garden • Woodland • Ponds • Permaculture demonstration plot • Community orchard • Tree nursery • Children's area • Organic vegetable allotments and herb project.

YOU CAN DO

Volunteering – in the grounds and on the waste-wood project.

COURSES

Composting for All This course enables participants to practise good composting for improved gardening and waste reduction.

Introduction to Green Woodworking Learn how to use green woodworking tools such as axes, froes, drawknives, spokeshavers, scorps, rounding planes, shave horses and travishers to make items of beauty and usefulness using locally sourced and fresh timbers such as ash.

Living Willow Structures

Permaculture Design

100. Greenspec

GreenSpec

Tel: 01733 238148. BrianSpecMan@aol.com
www.greenspec.co.uk

Directions Provided with course information (GreenSpec is not shown on the maps on pages 17-21, as courses are run all over the UK).

Description Greenspec Supports designers' and specifiers' everyday work with real solutions, and is developing a free-access website with really useful information for designers, specifiers, the construction industry, self-builders and the general public, including specification clauses for green building products. They present papers at green building events and display at exhibitions, run in-depth workshops on more specialised subjects and support government agencies in delivering sustainable construction strategies and initiatives.

COURSES

SWMP (Site Waste Management Plan) workshops • In-house CPD seminars – 340+ seminars on 30+ subject areas • Bespoke seminars or training to suit requirements.

101. Plants For A Future – Cornwall Site

 1400 DIFFERENT SPECIES

Plants For A Future, The Field, St Veep, Lostwithiel, Cornwall, PL22 0QJ.
Tel: 01208 873554. admin@pfaf.org or kenfern1@btinternet.com
www.pfaf.org/cornwall/index.php

Open Not open to general public except on specified Sundays – see website.

Directions Contact for directions.

Description Plants For A Future is a resource centre for rare and unusual plants, particularly those which have edible, medicinal or other uses. They practise Vegan-organic permaculture, with an emphasis on creating an ecologically sustainable environment based largely on perennial plants. In Cornwall, 'The Field' is Plants For A Future's original site, which was acquired in 1989. It is approximately 28 acres, and much of it has been planted with woodland trees and many other interesting plants. The woodland garden is 12 years old. Facilities at the site are very basic: there are no mains connections, so water comes from filtered rainwater, and electricity from a small wind generator.

Yearly event or point of interest The Field Experience in August – see how volunteers are restoring the site; listen to the many birds and music; taste some of the unusual edible plants grown; smell the pine trees; explore the gardens and learn about our solutions for a healthier world.

YOU CAN SEE (BY APPOINTMENT OR ON OPEN DAYS)

See and taste many of the unusual food plants grown, see many other interesting plants, and learn more about how and why we grow them. Just twenty plants provide the majority of food eaten, yet there are thousands of other useful plants which have not reached mainstream attention. Details of many of them are given at the site or on the website.

At the end, people are offered herbal teas and a special PFAF salad made from many of the food plants on the land. There is also the opportunity to ask for gardening advice and to purchase books, information sheets and plants. On many of the Sundays there is music and singing from one of the members of Plants for a Future.

YOU CAN DO

Volunteering. For details ring Addy on 01208 873554 • Work parties and music evening: to lead up to the open day a number of work-parties are organised – each one starts at 2pm and is followed by a Music Evening at 8pm. All are welcome to attend the work-parties, for details phone Phil on 01208 873623.

DEVON

102. Abey Smallcombe

Abey Smallcombe

Burrow Farm, Drewsteignton, Exeter, EX6 6PT.

Tel: 01647 281282. jackie@abeysmallcombe.com www.abeysmallcombe.com

Directions Private house. Directions sent out with course details.

Description Cob builders / designers who have been working with cob for the past 10 years.

YOU CAN SEE

Modern earth building • Cob • Mud-block building.

COURSES

A Practical Guide to Cob Building and Cob Repairs Our courses take place on a farm within the Dartmoor National Park. Each course covers the practical skills of mixing and building with cob. The courses last for one day, and lunch is provided. See website for future course dates.

103. Agroforestry Research Trust

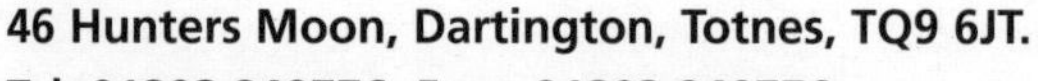

46 Hunters Moon, Dartington, Totnes, TQ9 6JT.

Tel: 01803 840776, Fax: 01803 840776.

mail@agroforestry.co.uk www.agroforestry.co.uk

Open Appointments can be made to visit, tours available.

Yearly event or point of interest We have several guided tours of our sites between spring and autumn.

Directions Contact for directions.

Description The ART undertakes research in temperate agroforestry, concentrating on unusual fruit and nut crops and on our 2-acre forest garden. As well as running courses it also sell plants, seeds, books etc.

COURSES

Crops for Climate Change This course looks at how climate change will affect (and is already affecting) how and what we grow, mostly in terms of tree and shrub crops. The most likely changes to climate (particularly in the UK) over the next 45 years will be outlined, and strategies for coping with drought, heat-waves etc. discussed, including irrigation systems. Novel crops likely to do well are also discussed. Includes two field trips.

Forest gardening This 2 day course gives an overview of how to design, implement and maintain a temperate forest garden. Teaching sessions are interspersed with frequent visits into our 12-year-old established forest garden.

Growing Nut Crops This course covers all aspects of growing common and uncommon nut crops in Britain. Teaching sessions are interspersed with visits to the forest garden and trials site where several nut crops are grown. Several unusual nut crops are also available to taste.

104. Centre for Contemporary Art and the Natural World

      OD TUE-SUN AND BANK HOLIDAYS, EXCL. CHRISTMAS

Haldon Forest Park, Exeter, EX6 7XR. Tel: 01392 832277.

info@ccanw.co.uk www.ccanw.co.uk

Open 10am-5pm.

Entry Free (there is a charge for events and activities).

Yearly event or point of interest In May 2007 CCANW is launching a year-long 'Wood Culture' Festival to highlight best practice in contemporary timber design and architecture.

Directions CCANW's Project Space is located at the heart of Haldon Forest Park, between Exeter Racecourse and Haldon Belvedere, 5 miles south of Exeter. Accessed off the A38 or A380.

Description CCANW is an innovative educational charity, focused on exploring our changing relationship to nature through the arts. It does so through exhibitions, residencies, projects, workshops, talks, and other events and activities. In April 2006 it opened a Project Space, the first phase of an exciting new partnership with the Forestry Commission in Haldon Forest Park.

The Forestry Commission provides trails for walking, cycling and horse riding as well as other activities – see www.haldonforestpark.org.uk.

COURSES

CCANW organises year-round workshops in its project space and the surrounding forest for adults and children. These relate to its changing programme of exhibitions, artist residencies and other projects.

In relation to 'Wood Culture', four exhibitions will provide the nucleus for a series of workshops and seminars that will investigate the use of timber in construction.

In Spring 2007 it organised the first forum to launch the University of the Trees. This long-term project is being developed by the Social Sculpture Research Unit at Oxford Brookes University in collaboration with other groups and networks through CCANW.

Details of the programme are published four times a year – see website. If you want to have the programme sent to you, contact CCANW.

105. Schumacher College

Schumacher
COLLEGE
An International Centre for Ecological Studies

The Old Postern, Dartington, Totnes,
Devon, TQ9 6EA.

Tel: 01803 865934, Fax: 01803 866899.

admin@schumachercollege.org.uk www.schumachercollege.org.uk

Open By arrangement. Courses need to be pre-booked.

Yearly event or point of interest Courses run all year except December and August.

Directions **Train**: 2 miles from Totnes railway station – possible to walk/cycle from the station along a Sustrans route. **Car**: 4 miles off the A38 between Exeter and Plymouth. Leave A38 on the A384, turn left by Dartington Church. The College is the next building on the left.

Description Schumacher College is an international centre educating about and inspiring sustainability. It runs short residential courses and a one-year MSc in Holistic Science for people of all ages and nationalities. Courses are taught by leading thinkers and activists such as Vandana Shiva and Fritjof Capra and address new economics, ecodesign, holistic science, education for sustainability, ecospirituality and new models of business.

COURSES

Climate Change: Seeing the Whole Picture (1/2/3 week options) This course will look at the whole picture and explore both the causes and consequences of climate change. It will show how innovative political solutions at the global level must work hand in hand with local action to deliver a clean and green form of prosperity. *Masters credits available.*

Creative Partnerships: Unleashing Collaborative Power in the Workplace (1 week) There is a growing recognition that in order to develop sustainable societies, the organisations running those societies must develop operating principles which support human dignity, peaceful interactions, and both economic and environmental viability. Organisations of all types – corporations, community groups, public agencies or religious and educational institutions – provide the context for many of our most consuming and productive relationships. Course participants will explore: how to transition from 'power-over' to 'power-with' relationships, what it means to be a 'partnering leader' and a new 'common ground model' which illumines and transforms conflicts.

Designing with Nature: Forms of the Wholeness (1/2/3 week options) Design is a ubiquitous and integral part of life: it can be the root cause of many complex problems or can become a powerful tool in solving them. This course explores how the principles of living systems can be applied to the design of landscape, food production, energy systems and even housing. *Masters credits available.*

Earth and the Sacred: The Personal and the Planetary (1/2/3 week options) Much of today's environmental thinking is based upon a predominantly scientific analysis of complex planetary problems. This course will explore the sacred, religious and spiritual dimensions of environmentalism to develop a vision where the care of the earth plays a central role. *Masters credits available.*

Illness to Wellness: Integrative Healthcare in the Community (1 week) Seventy percent of people in the Western world are diagnosed with some form of chronic disease, from high blood pressure to obesity. This course explores how contemporary healthcare can be greatly improved by incorporating the medical insights developed over millennia by our ancestors. The course tutors will work with participants to envision a new concept of healthcare which is local, holistic, integrated and preventative. They will explore whether an integrated model of local healthcare is realistic in the age of the quick fix and market competition. The course will provide insights into how to design a way of life which maintains wellness rather than merely treating illness. Based on their long experience of working in the mainstream, the course tutors will discuss how practical changes in the National Health Service can ensure that the future of healthcare is based on the quality of life as a whole.

Indigenous Peoples and the Natural World: Is Ancient Wisdom Relevant to the Modern World? (1/2/3 week options) This course will explore the relevance of the wisdom embodied by indigenous peoples all over the world. Course tutors will discuss the assaults being inflicted by globalisation upon indigenous peoples and their implications for planetary well-being.

Life after Oil: Breaking the Habit (two weeks) Many geologists, civil servants and industry experts are united in the view that the day the oil wells run dry is a lot closer than some think. Questions such as how did we become addicted to oil, and how can we redesign our society to free ourselves from oil dependence will be examined and discussed in this course. There will be an opportunity to imagine society, economy and culture in a life beyond oil dependency.

Roots of Learning: Education as a Springboard for Transformation (1 week) Greater instability in the Earth's ecosystems calls for a different type of education that challenges current lifestyles and values based on economic growth and consumerism. This is a timely short course for teachers or educators. The course teaches education for sustainable living and aims to give educators new teaching and learning skills with an ecological and holistic worldview. It is sometimes forgotten that the purpose of education is to teach the leaders of the future. And in this course, facilitators will introduce 'future orientated' concepts and models of sustainable education to participants. Experiential methods will engage participants in a spectrum of possible future scenarios, so that they can explore means of transforming the current educational system into one that is fit for the future with all its climatic and ecological implications. Participants will be able to develop ideas to meet the environmental challenges facing the educational system today.

Science and Spirituality: Creating a New Balance (1/2/3 week options) Can science help us appreciate beauty, explore consciousness and live a better life? This course will investigate the compatibility of science and spirituality and ask whether they complement or contradict each other. *Masters credits available.*

Solstice Retreat: Community, Symbolism and ritual (1 week) Take a break for a quiet Solstice Retreat. Enjoy the beautiful, tranquil surroundings of the College in the Devon countryside, curl up by the fire with a book, enjoy delicious organic vegetarian food, and take your pick from a range of optional activities. A variety of activities will be offered reflecting the theme of the season. Participants can join in discussions on the symbolism of the solstice, explore the night sky, create their own works of art, and take part in a Solstice Eve ritual centred on the College's labyrinth.

The New Economics: From Growth to Wellbeing (1/2/3 week options) Modern economic development has sought control over nature and material abundance through science and economic growth. This model has produced unanticipated consequences such as ecological destruction and the disintegration of local communities. The course will explore the creation of a new paradigm that puts well-being of people and planet at its core. *Masters credits available.*

106. J & J Sharpe

Furzedon, Merton, Okehampton, Devon, EX20 3DS. Tel: 01805 603587, Fax: 01805 603587. mail@jjsharpe.co.uk www.jjsharpe.co.uk

Directions The office is in the small village of Merton, about a 25-minute drive (12 miles) from Okehampton. Lime products are produced near North Tawton close to the A30, about 30 minutes from Exeter. Products can also be bought from Shiptons Builders Merchants in North Tawton. Please contact for location for courses.

Description J & J Sharpe are the longest established producers of lime products in Devon. They repair, refurbish and make sympathetic alterations to old buildings. Specialists in the conservation of early ornamental plasterwork, they also conserve and clean wall paintings and church monuments; promote and teach the use of traditional building materials, and carry out cob house renovation and repair. They are specialists in lime mortar, plaster and limewash and believe it is important to maintain our vernacular buildings and homes. Their aim is to employ the best techniques to achieve lasting repairs and maintenance of traditional buildings.

COURSES

The aim of the workshops is to give course members knowledge and understanding of cob building, cob repair and uses of lime in building, to enable them to undertake work themselves or to instruct contractors with confidence.

The cob and lime days are a mix of practical instruction and hands-on experience.

Cob Days Introductory talk and slides. Suitable soils. What soil type is required for successful cob building, and how can you find out? Mixing cob with straw. Making cob blocks. Building a mass cob structure. Mixing sieved cob for floors. Repair techniques. Use of stainless steel fixings, stitching and tools required. Appropriate renders, plasters and finishes.

Lime Days Introductory talk and slides. Choosing the correct sand, types of lime and hair for mortar. Health and safety, correct storage and shelf life of lime putty mortar and NHL. Slaking lime and mixing mortars. Pointing. Rendering techniques, hand harling and plastering. Protection and curing. Mixing limewash and limewashing.

107. Slapton Ley Field Centre

  6 WORKROOMS

Slapton, Kingsbridge, Devon, TQ7 2QP. Tel: 01548 580466, Fax: 01548 580123. enquiries.sl@field-studies-council.org www.field-studies-council.org

Open Not open to general public.

Accommodation 85 beds.

Directions 13 miles from Totnes station, 20 miles from A38. Please call for details.

Description See 'Field Studies Council' in 'National Organisations' at the back of the book.

COURSES

Badgers, Otters and Dormice Many of Britain's mammals are either rare or just plain difficult to find! Slapton Ley National Nature Reserve and the surrounding areas provide rich habitats for many of Britain's mammals, including otter, badger, and dormouse. This course shows you where and what to look for, and helps you to discover more about mammal ecology and conservation. There will be opportunities to see a wide range of species and to study tracks, trails and signs.

British Bat Ecology and Conservation Britain has 17 species of bat. This course provides an opportunity to practise identification skills through a variety of methods, including visual identification of bats in the hand and in the roosts, and through the use of bat detectors. It looks in detail at the ecological requirements of bats and a range of conservation issues including roost management and the importance of national roost monitoring, legislation and mitigation. Students visit a number of local roosts and consider the habitat requirements for foraging and flight paths, and discuss recent bat research.

Family Naturalists This course is for adults and younger family members who would like to discover more about natural history, and enjoy, record and photograph a wealth of natural history in the company of local experts at a variety of habitats: shingle beach and Slapton Ley National Nature Reserve, rivers, tors and woods of Dartmoor, mudflats of Kingsbridge Estuary, rocky cliffs at Start Point with nearby coves and pools, saltmarshes at Aveton Gifford and sand dunes at Bantham.

Green Lanes and Inns A daily walk is combined with lunch taken at a village inn. Travel through spectacular landscapes and explore the diverse local plant and bird life, while considering the geology, environmental issues and history of the area.

Using a Flora Particularly useful to countryside professionals, this course is intended for those with some knowledge of botanical terminology. It is designed for anyone who would like to increase their confidence in working with a flora as a means of identifying and getting to know more about our wonderful yet dwindling heritage of wild flowers.

Wild Weekend for Families This course aims to bring a sense of awe and wonder through a 'soft-focus' study of the natural world. This series of dynamic encounters with the natural world around Slapton is a combination of naturalist activity, arts interpretation, sensory awareness, bushcraft and traditional storytelling.

Wintering Wildfowl and Waders of South Devon This course takes advantage of Slapton Ley National Nature Reserve and the Exe and Kingsbridge Estuaries – all very different in character. Wader and wildfowl species will be building up to winter peaks. Avocet, widgeon and Brent geese should be on the Exe Estuary; greenshank and little egret on the Kingsbridge Estuary, and ducks such as goldeneye, pochard and tufted duck can be expected at Slapton Ley.

Woodland Plants The woodlands of South Devon should be at their richest in May – a treasure trove of plants to study and identify, from carpets of spring flowers to the graceful canopies of deciduous trees, freshly unfurled ferns and mosses in prime condition.

Many more courses on the website, e.g. **Plant Identification**.

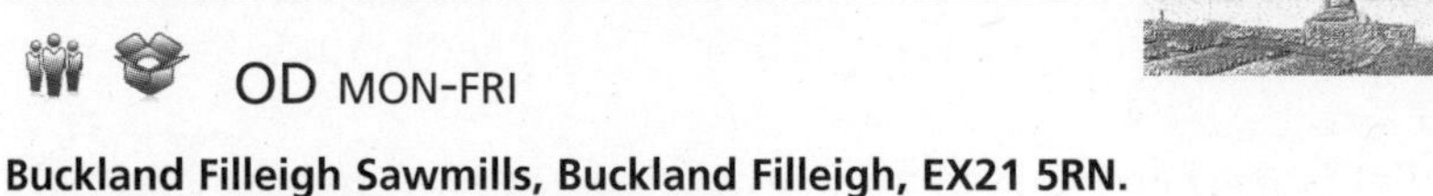

108. Mike Wye & Associates

 OD MON-FRI

Buckland Filleigh Sawmills, Buckland Filleigh, EX21 5RN.
Tel: 01409 281644, Fax: 01409 281669.
sales@mikewye.co.uk www.mikewye.co.uk

Open 8am-1pm and 1.30pm-4.30pm.

Directions North on A386 from Okehampton, west on A3072 from Hatherleigh for 6 miles, north through Black Torrington for about 4 miles.

Description Ecological builders' merchants, supplying natural building and decorating materials throughout the UK on a daily basis, from letter post up to 24-tonne lorryloads. Manufacturers of traditional building limes and lime-washes. Practical training courses in using 'lime in renovation' and 'polished plaster'. Building work undertaken. Consultations for building repair work.

COURSES

Using Lime in Renovation – General A relaxed environment with ample opportunity for discussion in the breaks and limited course numbers to ensure individual attention by two course leaders. For each session there is a practical demonstration, then participants have a go themselves. All tools and materials are provided, but please bring protective clothing, safety glasses, wellington boots and waterproof gloves.

Practical Lime Gain hands-on experience in using traditional lime and clay materials and techniques to repair and maintain old buildings. Course includes: traditional lime materials; building and repair in stone and cob; mortars; repair techniques; pointing techniques; protecting the work; soils; preparing the surface; rendering and plastering techniques; final finishes.

Venetian Plaster As a manufacturer and supplier of Venetian plaster, Marmorino and Tadelakt this course shows how to achieve fantastic finishes, including polished plaster and textured finishes. Traditional Venetian and Moroccan plasters: history and usage, types of finish tools used, health and safety. Types of surface and priming: applying the base coat of lime wall finish. Applying the first coat of Venetian plaster and Marmorino. Applying the second coat of Venetian plaster and burnishing to a shine. Marmorino: types of surface and priming, applying the base coat of lime wall finish, applying Marmorino and types of finish. Tadelakt: types of surface and priming, applying a base coat of lime wall finish, applying Tadelakt and polishing, applying polishing soap and waxes.

109. The Yarner Trust

    WILLOW BEDS

Welcombe Barton, Bideford, EX39 6HF. Tel: 01288 331692.
lun@yarnertrust.org www.yarnertrust.org

Yearly event or point of interest Summer Camp – an environmental arts holiday for families. Creative Breaks for 2007 – A Weekend in the Wood and Autumn Riches.

Directions Travelling from Bideford, turn right off the A39 Atlantic Highway at Welcombe Cross. Travel for one mile and take the right fork signposted to 'Church'. Travel for 1 mile, and when you see a stone dome on the verge, turn left into the Yarner Trust car park.

Description The Yarner Trust runs courses in sustainable and creative living from a medieval farmhouse recently featured in BBC2's Restoration Village. It is renowned for its sustainable buildings programme, from straw-bale to green oak; living in a sustainable home to rainwater harvesting; inspirational textile, mosaics and willow weaving.

YOU CAN SEE (BY APPOINTMENT)

Straw-bale building • Cob building • Timber-framed building • Renewable insulation • Compost toilet • Rainwater collection.

YOU CAN DO

Volunteering.

COURSES

A Feast of Christmas Choose two of the following activities: Swags and Garlands, Textile Decoration and Cards, Unique Willow Christmas Trees, and Music and Singing. Complete the day with mulled wine, mince pies and a musical performance.

Backyard Poultry Discover all aspects of keeping poultry and water fowl in your back garden: breeds, housing, feeding, handling and caring for birds on a small scale. Includes a local site visit.

Baskets from the Beach Winter storms wash in a tangle of variously coloured ropes, nets and driftwood on to the shore at Welcombe Mouth. Using Somerset willow, shells and other beachcombings, create your own unique basket.

Bicycle Baskets Create a sturdy basket on an oval base with leather straps to attach to your handlebars. Basic willow basketry skills needed.

Bronze Casting Under the guidance of sculptor Holger Lonze, you will design and make sculpture and jewellery objects in bronze in this practical workshop, using the lost wax process and carbon-neutral pit furnaces, together with locally abundant and regenerative materials.

Cleft Post and Rail Fencing Learn to make cleft post and rail fencing using locally sourced oak and sweet chestnut.

Creative Knitting Experiment with stitches, colour and texture, and enjoy a totally new creative approach to this textile craft.

Creative Textiles Create your own fabric using Angelina – a fibre which when heated creates unusual colour and textural effects. Add a few embroidery stitches, snippets of material and some recycled threads and you can create a unique pin cushion or book cover.

Cultivation and Use of Willow Visit our organic willow beds, choose cuttings from our specially selected varieties, and complete the day with a simple woven project to take home and plant.

Devon Stave Baskets Make a basket with cleft and bent ash. A knowledge of some practical hand tool skills would be an advantage.

Dry Stone Walls Learn dry stone walling using tried and tested traditional methods. This practically-based day introduces you to the techniques, use of tools, and aesthetic and wildlife benefits of the traditional stone wall.

Edible gifts Create mouthwatering and individual gifts, including herb oils and vinegars and delicious hand-made chocolates, and discover some original packaging ideas.

Frame Baskets Make a small, rustic, traditional basket using locally produced willow of various hues. Ideal for both beginners and experienced makers.

Fungal Foray Learn the characteristics that distinguish species of wild mushrooms from toadstools, and then join us on a fungal foray. Learn about fungal ecology and, with luck, sample the best edible species.

Green Oak Building Work with a master craftsman on our small green oak eco-build project. Learn the techniques associated with timber-framed building – making oak pegs, using naturally grown curves, and mortice and tenon joints.

Jigs, Reels and Shanties explore and enjoy traditional music in a relaxed and supported atmosphere whether you are a beginner or an intermediate player.

Kumihimo This is a traditional braid-making technique which has been an integral part of Japanese culture for many centuries. Learn how to make and use a *marudai*. Combine wrapping, tassel-making and braid-making to develop a range of unique embellishments for creative textile projects.

Lime rendering A practically-based workshop which covers pointing, rendering, plastering and limewashing.

Living in a Sustainable House Looking at individual lifestyles, learn how to calculate and reduce your carbon emissions and conserve energy and water. Visit local examples of solar and wind power, and discuss the suitability for your home. A practical interactive course providing economic and ecological advantage.

Living Willow Structures Arches and bowers, trellises, fences and screening, playhouses and tunnels are just some of the projects that can be made using simple weaving techniques. Design one for your space and learn the practical techniques by creating a structure *in situ*.

Machine Embroidery Discover the potential of your sewing machine. Straight stitch and zig-zag are enough to create beautifully decorated fabrics which can be transformed into a tasselled cushion cover, a folded wallet or a bag.

Mosaics from Old Masters Create a mosaic drawing on the work of your favourite artist as a source of inspiration using recycled china, glass, beads etc in addition to traditional materials to embellish the design.

Off the Sheep's Back Spend a day on a working smallholding at lambing time to look at the management cycle of a small sheep flock; then learn how to process the fleece using either a drop spindle, spinning wheel, or felt making.

Orchards and Apples Spend the morning with George from Westlake Orchards, learning about orchard management – from selection of varieties to pruning. Then, press and take home your own freshly-made apple juice and discover some of the techniques for successful cider and vinegar making.

Rustic Furniture Making Anyone can make an individually designed rustic chair or table using coppiced wood. This course explores design, main construction methods and styles of rustic furniture and develops a knowledge of traditional hand tools.

Smallholding and Self-Sufficiency Take a closer look at the ideals and realities of a self-reliant lifestyle on a few acres of land. The course covers sustainable production of food from land and livestock, basic skills needed and the potential for earned income. Visits to established smallholdings provide working examples. Alan and Rosie run a 16-acre smallholding along organic lines and write regularly for *Organic Smallholding* magazine.

Spoon Carving Using a simple selection of tools, learn the carving techniques required to create unique utensils from locally sourced green wood.

Spring into Song A lively and fun day exploring singing in a variety of styles. No experience necessary.

Straw-Bale Building This practically-based course covers all the building principles and skills necessary to understand this sustainable building material.

Stripped-Willow Furniture Making Make a simple but beautiful piece of furniture from stripped willow. Explore the design and construction possibilities of making a table or chair, combining a stripped willow framework with multi-coloured withies from our own beds.

Traditional Rakes and Besom Brooms Learn to make and take home your own versatile and efficient garden tools.

Traditional Soap Making Create your own fragrant soaps using natural locally sourced ingredients. Learn something of the history and chemistry, and take home your own samples of delicate lavender, scrubby gardener's soap and a refreshing seaweed soap.

Using Natural Dye Plants Natural dyes from all over the world can be used to achieve a rainbow of colours on wool and silk. The day covers the preparation of fibres for dyeing, the equipment needed, harvesting the plant material, dyeing and then modifying the results to create a whole spectrum of colour. Fresh dye material will be collected from our own dye garden where possible.

Wattle Hurdles Learn the techniques of splitting, twisting and building up the weave, and take home a small hurdle.

Willow Backpacks Using both Somerset and locally grown willows, create a functional and attractive backpack basket for gathering and storing kindling or driftwood. Learn the techniques of D-shaped frame basis, scalloms and slewing. Some experience would be useful.

Willow Trugs Combine local and Somerset willows to create this attractive and robust trug, ideal for gathering your garden produce or wild fungi. Some experience would be useful.

Woodcuts and Print Making Learn the process of producing successful black and white wood cuts from the initial concept, through the individuality of the block itself, to the print. Experiment with cutting techniques and then print by hand on a range of papers.

Working with Cob Spend the day working with cob in a traditional repair / rendering and creative context. Make cob lifts for repairs to existing buildings, and create a decorative seat or BBQ feature for the garden.

Yarner Week An environmental camping arts holiday with a range of workshops, storytelling, theatre events, barn dances. Ring for separate leaflet.

Yurt Making Learn the skills and make some of the components for your own 12-ft yurt frame.

110. Upcott Project

    OD MONDAY-FRIDAY

Lower Upcott Farm, Hatherleigh, Okehampton, EX20 3LN. Tel: 01837 811123. ben@forestcrafts.co.uk www.forestcrafts.co.uk

Open 10am-4.30pm.

Entry Free. Group tours (more than 6 people) need to be pre-booked.

Accommodation Holiday let.

Directions **Car** From the roundabout on the A386 just outside Hatherleigh in Devon, turn off following signpost to Town Centre. Take the second turning right signposted Monkokehampton. Follow road uphill for 1 mile on to Hatherleigh Moor. You will pass a monument on the left. Take the next turning right, signposted Upcott. From here, Lower Upcott Farm is 1.2 miles and is the second farm on the left.

Description At the Upcott Project, inspiration comes from wood and livelihoods depend upon it. Visit the farm and share the vision of a sustainable future for Britain's woodlands and countryside. Lower Upcott Farm comprises around 43 acres of woodland and pasture, both of which are worked by shire horses.

YOU CAN SEE

Cob building.

COURSES

Basket Making for Beginners Introduction to basic basket making techniques, materials and tools required. Make your own basket.

Charcoal Burning Charcoal burning is an ancient craft where a high quality and high value product is made from low value and low quality timber. This course takes you through the whole charcoal-burning process.

Hurdle Making and Coppice Management Take a trip to the woods to learn about sourcing hazel material and how to manage coppiced woods. Then make your own hurdle to take home.

Rustic Furniture Making We shall explore the main construction methods and styles of rustic furniture making. Using a range of traditional hand tools and devices, you will construct your own chair to your own unique design. We very often find that your character comes out in the design of the chair that you make.

Tailor-made courses Groups wishing to come for a weekend or longer will be able to choose a number of courses to suit their interests. For large groups, we shall be able to give a choice of courses for each day to suit different interests. Accommodation may be possible in the holiday let.

Working a heavy horse (1 or 2 day, 1 or 2 people) This course provides as much 'hands-on experience' as possible of working a heavy horse and is tailor-made to your requirements.

CHILDREN'S ACTIVITIES

Kids' Indian Tribal Nature Day (age range 7-12) Activities include a horse wagon ride, bug hunting, a treasure hunt, simple compass work, shelter building for survival and communication exercises based on an environmental theme. Parents are welcome to stay and help if they wish.

In addition, a wagon ride is provided for children's parties, and further entertainment for children in the form of educational farm and woodland games and shelter buildings can be provided.

DORSET

111. The Cherry Wood Project

OD

Whistley House, Milton on Stour, Gillingham, SP8 5PU. Tel: 07921 361484.
info@cherrywoodproject.co.uk www.cherrywoodproject.co.uk

Yearly event or point of interest Full programme of courses on the website. Appointments can be made for tours.

Directions The Project is located in a 40-acre wood, 6 miles north-east of the City of Bath. Contact for full directions.

Description The Cherry Wood Project runs a range of courses to teach a variety of woodland skills.

YOU CAN DO

Volunteering.

COURSES

Advanced Green Woodwork (9 days) This course covers most chair-making techniques including advanced lathe work, complex steam-bending and chair assembly. Skills include: adze, travisher, inshave and spokeshave, advanced steam-bending, bark preparation and weaving. There will be time to complete a demanding project such as a chair with arms.

Basic Bushcraft Learn and practise a range of survival skills; test your shelter-building ability; try a range of equipment; take advantage of the abundant wild food produced at the beginning and end of the year – eat well. Skills introduced: firelighting without matches; cooking and fire management; foraging for and preparing wild food; shelter building; water purification.

Basic Green Woodwork (2 days) A course designed to give students practical experience at the shave horse, pole lathe and a variety of green woodwork tools and techniques. No previous experience is needed.

Build a pole lathe / shave horse (2 days) All the equipment instruction and materials you need to build your own pole lathe or shave horse. Skills: timber selection, cleaving and axe work, use of hand tools, pole lathe.

Coppice weekend A chance for the whole family to learn the skills of coppicing and restore an area of neglected hazel and field maple coppice. The days are hard work and the weather may be harsh, but there will be a roaring fire and a warming lunch! Skills introduced: use of hand tools, billhook, saws and axes; formation of coppice stools; dead hedging; selection of coppice materials.

Earth Oven (2 days) Learn how to build an earth oven from the soil in your garden.

Harvest and Weave Elm Bark Learn how to harvest the bark of the elm tree. Weave containers and chair seats. Skills introduced: Identification and selection of elm trees; harvesting, preserving and weaving bark.

Intermediate Green Woodwork (6 days) A course for students with a little woodworking experience who want to tackle a more complex project. Skills include: use of the froe; draw knife and shave-horse; use of adze, travisher and inshave; simple steam-bending; preparation and use of elm bark; practical sharpening.

Introduction to Green Woodwork (1 day) The day is designed to give students hands-on experience of the variety of tools and skills used in green woodwork. Selection of timber; review of green woodwork and hand-tools; cleaving and axe work; draw knife and shave-horse; pole lathe; knives and carving tools.

Tool Sharpening and Restoration (1 day) This day teaches you how to select hand tools, new and second-hand, achieve and maintain a professional edge and select, restore and repair old tools. Includes sharpening skills for knives: draw knife, axes / adzes, chisels / gouges, planes / spoke shaves, travishers / spoon knives. It is essential that you bring your own tools.

112. Dorset Centre for Rural Skills (DCRS)

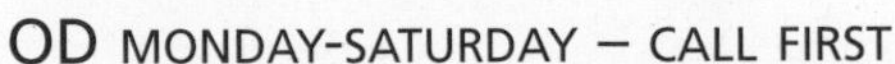
OD MONDAY-SATURDAY – CALL FIRST

West Farm Barn, West Farm, Farrington, Nr Blandford, DT11 8RA.
Tel: 01747 811099, Fax: 01747 811078.
info@dorsetruralskills.co.uk www.dorsetruralskills.co.uk

Open Mon-Fri 8.30am-5pm, Sat 9am-12.30pm. Although the Centre welcomes visitors, it is a small Centre and it is advisable to call first.

Directions **Car:** From the A350 at Iwerne Minster take the Child Okeford Road. Take the first turning on the right (approx 1 mile) signposted to East/West Orchard. At end of road turn left. Enter Farrington, at end of road turn right and immediately right.

Description Dorset Centre for Rural Skills (DCRS) is an established training centre. Courses and events cover sustainable and traditional building such as straw-bale building and lime days; rural skills including dry-stone walling and hurdle making; also arts and crafts ranging from glassblowing to blacksmithing. The Centre is also home to the Sustainable Building Resource Centre, a straw-bale building that demonstrates many of the techniques used in sustainable construction. It is a one-stop shop for sustainable building, providing information and resources, advice and design facilities and a wide range of building materials and products.

YOU CAN SEE

Solar heating • Passive solar design • Information on different aspects of renewable energy technologies • Straw-bale building • Cob • Rammed earth • Timber-frame • Renewable insulation • Rainwater collection • Other sustainable living materials.

COURSES

Art Classes – Exploring Mediums Explore, experiment, discover and exploit a variety of art materials. Develop a sketchbook of wet and dry techniques, expand your visual vocabulary and understanding of materials.

Art Classes – Life Drawing Explore the rhythm, movement and feeling of the living form. Discover the language and landscape of the body through close observation. Study anatomy and the basic principles of working from a life model.

Blacksmithing Spend a day in a fully equipped forge. Course includes: learning about the workshop and practising basic blacksmithing techniques; health and safety; tools and equipment; and how to light and use the forge; basic metallurgy; working on an anvil and swage block, pointing, curling and forge welding. Put all these techniques into practice and make a small project to take home.

Cob and Earth Building A day of theory and practical work which includes history of cob; various types of cob and earth buildings; health and safety; good building practice; making cob, and the methods of building with it; production of cob blocks; rammed earth; repair and maintenance; use in new buildings; the role of cob and earth buildings as an environmentally friendly material.

Dowsing A dowsing course for beginners run by Miranda Aldridge of the Shaston Dowsing Group. The day consists of an introduction to dowsing. Participants will make their own dowsing tools (which can be taken home at the end of the day) and spend the day learning and practising dowsing techniques around the centre, including a local field location.

Glassblowing A day of glassblowing, learning the basics of how a glass studio operates, health and safety in the studio and how to use the equipment. After a demonstration there will be plenty of practical work suited to your own ability. Spend the day practising these techniques and end it with your own piece of work. This will need to spend 24 hours in the annealing oven and can be collected later.

Greenwood Furniture A 2-day practical course with Pete Moors. Working in hazel, learn the techniques of creating greenwood furniture by creating your own piece. Depending on ability, it is usual for students to make a chair over two days, which can be taken home at the end of the course.

Hedge-Laying A practical hands-on weekend run by Pete Moors of the Dorset Coppice Group, spent outdoors, regardless of the weather! It covers basic health and safety, tool use, sharpening and maintenance of tools and all methods of hedge-laying.

Hurdle Making and Coppice Crafts A 1-day course run by Pete Moors of the Dorset Coppice Group. The course covers health and safety, tool use and working with hazel.

Knots and Splicing A practical day with Pete Townsend in the main hall at DCRS learning about knots and splicing.

Lime – Principles and Practice The Lime Day is a blend of theory and practical work covering everything from health and safety basics, to applying renders, plasters and limewashes. The course covers all aspects of lime: history, production, types, uses, application on different surfaces, lime foundations, tools and their uses, lime mortars and the role of lime as an environmentally friendly material.

Pottery This eight-week course teaches all the processes of pottery, including: preparing the clay, throwing pots on the wheel, turning, coil pots, making handles and glazing. The tutor will fire all pots, but will explain the different drying and firing stages. Students finish the course with their own selection of pots and about half decide to continue with pottery having done this course.

Sculpture Classes Working with a life model on a series of exercises and studying the work of respected sculptors past and present. Subjects include wire drawings, working in relief, portraiture, and clay sketches and studies. Charlotte Moreton trained at the Frink School of Figurative Sculpture and combines sculpture and teaching with work in landscape design and ecology.

Straw-Bale Building A 2-day course of short lectures and practical building. Subjects include: history; advantages and disadvantages; planning and building control implications; environmental impact; costs; planning your project; foundations; walls; roofing; finishing options; maintenance.

Textiles and Weaving 2-day courses are run by Daisy Bewes of Weave World at DCRS.

Timber Framing Two days of theory and practical work which includes: woodland management and timber extraction under a 'constant cover' system; species of timber and their uses, advantages and disadvantages; working green wood; working with wood in the round; converting timber from the round and grading; design and structure of buildings; techniques and jointing methods; tools, their use and heath and safety; cutting, jointing and assembly of frames; breathing structures.

Using the Six Traditional English House Paints (2 courses) DCRS is hosting two courses on traditional house paints, both run by Nature Paint: **Water-based paints: washes, distempers and emulsions** and **Traditional oil-based paints: roman (or masonry) paints and glazes**. Both courses will cover a mixture of theory and practical work and at the end of each course, students take home any paints they have prepared on the day.

Welding (1-day or 6-week course) Learn the correct way to arc, mig and gas weld. The practical day course includes: health and safety; tools and equipment; basic metallurgy; metal, steel and non-ferrous; judging heat of metal by its colour; effect on metal of different temperatures; theory and practical use of arc, mig and gas welding; testing and dressing welds; the use of grinders and files. Most of our students can practise their new skill confidently at home. For students who do not have the opportunity to practise welding, our six-week course offers the chance. Beginners are welcome.

113. Monkton Wyld Court

Monkton Wyld Court, Bridport, Dorset, DT6 6DQ.

Tel: 01297 560342.

info@monktonwyldcourt.org www.monktonwyldcourt.org

Accommodation Organic B&B – up to 35 visitors.

Yearly event or point of interest Open Day usually in September – 25th anniversary in 2007.

Directions 4 miles from Charmouth, Axminster and Lyme Regis. The village of Monkton Wyld is signposted from the A35 halfway between Charmouth and Axminster. The Court is opposite the church.

Description Monkton Wyld Court is a holistic education centre run since 1982 by a resident community, with help from visiting volunteers. The main building is a beautiful Victorian rectory, set in 11 acres of grounds. A full programme of workshops throughout the year covers a wide range of subjects. Monkton Wyld can be hired as a venue for your own group or workshop.

YOU CAN DO

Volunteering.

COURSES

Body Mind Centering Exploration through movement of the immense potential of the body. We will awaken movement at its source within the different cells and tissues of the body, and respond to the body's primitive thirst for movement and meeting. Body Mind Centering (BMC) is an approach developed by Bonnie Bainbridge Cohen.

Celebrating Nature Aimed mainly at people working in education or with community groups, this workshop provides a practical introduction to the use of art as a way of reaching out and responding to the natural world. Experiment with stories and poems, make lanterns, populate the woods with puppets, make masks, and end the workshop with a portfolio of activities, some ready to use right away and others to develop further.

Community Work Week Join us for a week of hard work, sharing skills and learning new ones. Work varies from constructing our compost toilet, to enjoying some organic gardening. You will also experience the workings of the community through a little cooking, cleaning, etc. There are organised activities in the evenings, or stroll up the road for a pint, or take a trip to the seaside.

Dance in Body and Soul This workshop is an invitation to come home to your body through 5 Rhythms™ Ecstatic dance. The 5 Rhythms™ is dance as a wild, freeing, deeply physical meditation. Its journey begins in receptive connection to the self, builds in power and expression to a high energy ecstatic release. This gives rise to the effortless movement of vibrant, creative, soul energy and finally settles into the bliss of deep stillness within movement. "Put the psyche into motion and it will heal itself" – Gabrielle Roth.

Find your Power Introducing strategies from Chris's new book *Find your Power*, this course offers the tools to develop inner strength, clarity of focus, deepened determination and the ability to overcome obstacles.

Firebird Trance-Dance and Sweatlodge Ceremony In addition to the trance dance (see earlier entry), this workshop includes a sweatlodge, the ancient purifying healing ceremony.

Gaia's Playground – Exploring Clowning and Ecology The clown's exaggerated physical and emotional responsiveness gives us a heightened sensitivity to the world around us. Through the clown's empathy and playful imagination, we open ourselves to the wonder and mystery that is Gaia's playground. The weekend is an opportunity to let go of thinking, controlling part of ourselves and open up to the here and now. Clown beginners welcome.

Introduction to Green Woodworking A practical course introducing the theory and practice of working unseasoned timber. No prior knowledge or experience is necessary. The basic methods of cutting, cleaving, shaving and turning, using a cleaving break, shave horse and a pole lathe are demonstrated, enabling you to practise at your leisure. With skills honed, you can make something from a range of small projects. All tools are provided.

Home Educators' Development Week This week offers a time for home education families to meet, learn new skills, access resources and to play! The week includes: sharing knowledge: trying crafts such as felt making and pottery; making friendships; workshops run by representatives of Education Otherwise, the Home Education Advisory Service and Human Scale Education.

Kriya Yoga: Purification and Reflection Nearly 2,000 years ago, the sage Patanjali presented the concept of Kriya Yoga as a means to stability. This weekend workshop explores the relevance of his ideas today. Take time out for some stillness, reflection and discussion as well as posture work (*asana*), breath work (*pranayama*), chanting and meditation.

Permaculture Design Course Permaculture is a design system for sustainable living covering everything from organic food growing to greener building. This course is relevant to urban and rural situations, and includes practical and theoretical sessions. It follows the 90-hour syllabus, and is certified by the Permaculture Association (Britain).

Revitalise Your Life Through Creativity An inspirational women's weekend. Take time out to bring a higher awareness to your self-development through movement, drawing, writing, drama and sharing. Working with the edges of our comfort zones, within a supportive environment, we begin to understand more about our limits and the risks necessary to move forward. Focus is on the process and experience of the participant rather than tasks or outcomes.

Sacred Drumming Drumming is a spiritual experience, being with the rhythm of nature. Celebrate the sacred drummer in all of us. Doug teaches traditional rhythms of Guinea and Mali (West Africa), along with the history and the cultural context for each piece and the ethnic grouping the music comes from.

Sing Your Heart Out! Explore the full expressive potential of singing, beginning with gentle physical work to release accumulated habitual tensions, then working with whole body, doing breathing and voice exercises to enhance projection and develop vocal strength, range, pitch and clarity. Sing a wide range of songs in harmony, including world music, spiritual and traditional chants and rounds. No experience of singing or harmony is necessary.

The Courage to Sing This workshop is for people who want to access and increase their creativity and build their self-confidence. Focused on singing, it will also include breath work, jamming, performance, visualisation, process work and discussion. All are welcome and everybody will receive support to work at their own level.

The Ecology of Money Why is the US the world's sole superpower? Can we make sufficiently drastic cuts in greenhouse gas emissions to avert catastrophic climate change? The answers to these and other crucial questions turn on the types of money we decide to use in future and how those monies are put into circulation. This course looks at different sorts of monies – money created for taxation, a banking system for profit – and how they affect communities.

The Goddess and the Whore (exploring the sacred and sexy) Give your spirituality and sexuality space to flourish. In a safe and sacred space, using ritual and ceremony, we will support and witness each other, sharing and celebrating our journey, our stories and our gifts. The weekend also includes other forms of creative bodywork, massage, play, dress up and let your hair down!

The Shaman's Path An introduction to the world of the shaman. Shamanism is the oldest spiritual path of healing and keeping in balance with yourself, the living Earth and all creation. This workshop explores the world of the shaman and how it can be applied to the modern world, and includes experiential work outdoors exploring our connection to the natural world.

The Work that Reconnects Are you concerned about the condition of our world? Would you like to strengthen your ability to respond? This workshop introduces The Work That Reconnects, an empowerment approach developed by Joanna Macy (see www.joannamacy.net for info). Deepening our feeling of connection with life, we will open up spiritual and psychological resources needed to face and respond to global issues.

Trance-Dance at Equinox This workshop will move you profoundly towards the truth of yourself and your life. The Trance-Dance is a magical catalyst for shifting patterns and allowing inner deepening. It enables you to sweep away confusion and delusion and to penetrate to the heart of your energy and connect with your soul's purpose.

Wild Food on the Solstice Celebrate the Summer Solstice gathering and eating wild food in the wilds of Dorset. Learn to identify plants, share recipes and learn about preparation and preserving methods, and connect with the bigger picture of sustainability and our food culture. This is a chance to open your senses to the fundamental and ancient knowledge of the gatherer, to Nature and to Spirit.

Wonderful Willow Weaving Weave the basket of your choice using some of the beautiful colours of willow, or make a structure to enhance your summer garden, or a scarecrow or some simple decorations. Suitable for beginners and improvers. All materials and tools supplied by the tutor.

Writing to Reconnect: Seeking the Soul This weekend workshop combines writing and meditation to reconnect with trees, water, earth and sky, using haiku and other writing forms. Through creativity and imagery, we explore and nourish both our own soul and the collective soul of the world. Experience of writing and/or meditation is not necessary.

Wyld Winter Waves We will spark and nourish the fires of our inspiration, and drop deeply inside ourselves to prepare our inner ground for the coming of spring. Dance brings fluidity, flexibility and vitality to our bodies. This naturally warms and opens the heart, releasing the spontaneity of our dancing spirit.

FAMILY ACTIVITIES

Crazy Kids Family Week A relaxing week of summer fun. There will be an emphasis on music, singing and drama as well as the usual activities so if you have an instrument, please bring it along! Crazy kids from toddlers to teenagers are welcome.

Fireside Family Week This week focuses on storytelling and drama. The mornings are filled with interactive stories led by a storyteller. Meanwhile Ann will be teaching a range of colourful crafts, taking inspiration from the story sessions and the unique creativity of each one of us.

Funky Spring Family Week A seasonal Easter celebration, where you can enjoy a range of crafts and games, bushcraft, music making and singing. This week is suitable for families of all ages – bring an instrument if you wish.

Wyld Easter Family Week Celebrate spring and have fun through music, drama, art, craft and much more. Witness this time of new beginnings and enjoy nature in this beautiful environment. Taste community life by helping in our kitchen, walled garden, farm and grounds, or just relax and take it easy.

Mind and Body Family Week A week of fun craft activities, an outing to the beach, a picnic and walk to Lyme, a pebble fair, dance, music and story sessions, a bonfire and much more. Regular yoga and meditation sessions and space for all to relax and enjoy a refreshing week for the mind, body and spirit.

Midsummer Madness Family Week

Summer Surprise Family Week Play, explore, make stuff, get messy and experience our community. Celebrate summer, take time to learn, to rest and to get down to some serious sunny fun. You'll be joined by a team of experienced facilitators and creative instigators, here to support and inspire you during your stay. Contact us or see our website for more details.

114. Studland Education Centre

CLASSROOM

Purbeck Estate Office, Studland, Swanage, Dorset, BH19 3AX.

Tel: 01929 450259, Fax: 01929 450508. studlandbeach@nationaltrust.org.uk

Open Please call.

Directions See website for full list of directions

YOU CAN SEE

Compost toilets • Photovoltaic solar energy system • Sustainable building.

YOU CAN DO

Guided walks.

COURSES

Please contact for a list of events.

GLOUCESTERSHIRE

115. Ragmans Lane Farm

Ragmans Lane Farm, Lower Lydbrook, GL17 9PA.

Tel: 01594 860244, Fax: 01594 860244. info@ragmans.co.uk www.ragmans.co.uk

Accommodation Eco-hostel, camping, bunkhouse.

Yearly event or point of interest Courses throughout the year – see website. Open day October 2007.

Appointments Farm tours are £20. Please pre-book all visits.

Directions **Bus** from Gloucester: the bus station is next to the train station. Get a bus from Gloucester to Ruardean. Buses are Cotterels (purple) (01594 542224) or Beaumont Travel (white) and they leave on the hour; the last one Mon to Sat is at 8.00pm. Ruardean is only ten minutes walk, or you can phone us. See website for directions by car.

Description Ragmans Lane is a 60-acre farm in the Forest of Dean in Gloucestershire. It produces shiitake mushrooms, willow, comfrey and apple juice. It also teaches a range of courses from the farm throughout the year. Although it is a small farm, a diverse mix of enterprises provides employment for four people. The farm is run on permaculture principles and is primarily about educating and employing people to work the land sustainably, giving them the opportunity to 'learn on the job'.

YOU CAN SEE (BY APPOINTMENT)

Biomass production – pellet • Biomass production – willow • Biofuel • Exchange heat pump • Solar heating • Tile stove • Passive solar design • Wind turbine – domestic in 2007 • Cob building • Timber-framed building • Renewable insulation • Compost toilet • Rainwater collection.

COURSES

Alternative Energy Solutions for the Home • Cider and Perry Making • Jam, Marmalade and Chutney • Organic Gardening for Beginners • Permaculture Design Course • Practical Applications of Herbalism • Restorative Fruit Tree Pruning • Sustainable Land Use • Willow Coppicing • Willow Weaving.

116. The Green Shop

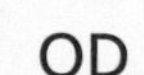 OD

Cheltenham Road, Bisley, Stroud, GL6 7BX. Tel: 01452 770629, Fax: 01452 770104.

enquiries@greenshop.co.uk www.greenshop.co.uk

Open Mon-Fri 9am-7pm, Sat 9am-6pm, Sun 9am-1pm.

Directions Located behind Holbrook Garage one mile north of Bisley near Stroud in Gloucestershire. From the road, look out for the red canopy on the garage forecourt, and the solar panels on the roof.

Description The Green Shop encourages customers to live lightly on the planet by providing a wide range of sustainable and low-impact products for the home. The new eco building (opened in Spring 2006) provides an excellent backdrop, demonstrating many sustainable building techniques including solar thermal heating, photovoltaic solar panels, a rainwater harvesting system and a green roof planted with sedum. The interior is decorated with natural paints and finishes all available from The Green Shop.

YOU CAN SEE

Biomass • Biofuel • Photovoltaics • Solar heating • Passive solar design • Wind turbine – domestic • Timber-framed building • Renewable insulation.

117. The Organic Farm Shop

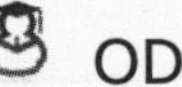 OD

The Organic Farm Shop, Abbey Home Farm, Burford Road, Cirencester, GL7 5HF. Tel: 01285 640441, Fax: 01285 644827.

info@theorganicfarmshop.co.uk www.theorganicfarmshop.co.uk

Open Tue-Sat.

Entry Free entry to the farm shop and café, and woodland walk. Please pre-book group farm tours, meeting room, and hire of yurts, eco camp, campsite, hut and holiday cottage.

Accommodation Camping, holiday cottage, single yurt and 4-yurt camp.

Yearly event or point of interest Ragged Hedge Fair – first weekend in September. Community fair powered solely by the wind and sun. Music, crafts, children's entertainments, eco forum, farm tours, magic, organic food and organic bar. Food and Nutrition Club at the farm shop once a month.

Directions Where the A417 crosses the B4425, follow the B4425 towards Burford. Go past the A429 and soon afterwards you will see a sign to turn left for the Organic Farm Shop.

Description Award-winning organic farm, farm shop and café. Growing and raising local food for local people in a sustainable way. Lots of ways to enjoy staying on the farm: school visits, day visits and residential at our eco camp and holiday cottage. Away days for businesses, schools, special needs groups, workshop/conference space to hire. Connecting people with where their food comes from.

YOU CAN SEE (BY APPOINTMENT)

Woodchip production • Biomass boiler • Solar heating • Compost toilet • Green oak building.

COURSES

We offer a range of cookery courses throughout the year for adults and children, always inspired by the produce on the farm; they take place in the purpose-built teaching kitchen. See website for more details.

A Vegetarian Christmas Preparing and cooking special / luxury foods that will freeze ready for Christmas.

Garden Vegetable Cooking 'Cooking with no food miles!' Making the most of the seasonal vegetables in the farm gardens.

Japanese Cooking Learn to make traditional Japanese food. Make vegan *maki sushi* (*nori sushi* rolls with rice) and vegetable *tempura*, and share the meal we have created together. Cost includes ingredients, recipes and a Japanese meal.

Relishes, Pickles and Sauces 'Tantalise your tastebuds – enliven any meal!' Create a colourful palette of dishes by hand from fresh garden produce and seasonings from around the world.

Seasonal Cooking and Simple Living Led by Hilary, who started the farm shop and garden project, and Dawn, who uses produce from the garden to cook in the café each day, this weekend residential course shows you how to cook seasonal vegetarian dishes using ingredients from the organic kitchen garden. Part of the weekend experience will be staying a night on the farm in one of the yurts (two sharing in each), in the hut by the pond or in the farmhouse.

CHILDREN'S ACTIVITIES

Cookery Class for Children, Ages 6-10 Visit the garden and explore all the senses – touch, smell, taste and hearing, and pick fresh vegetables to make a simple lunch to take home.

Kids Cookery for Ages 6-9 We go outside to see our chickens to find out where eggs really come from! Then back to the kitchen to prepare a simple autumnal lunch to take home. Cost includes ingredients and a snack.

Kids Cookery for Ages 9-14 Come and learn how to make a simple Halloween supper to impress your family and friends. You will take home recipes and the food we've prepared. A hands-on experience of cooking with no food miles!

118. Thistledown

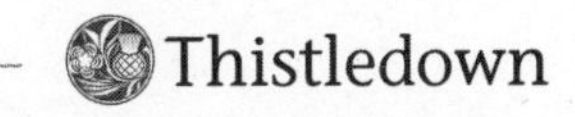

 OD

Tinkley Lane, Nympsfield, Nr Stroud, GL10 3UH. Tel: 01453 860420.

thistledown@toucansurf.com www.thistledown.org.uk

Open 7 days a week, 10am-6pm.

Entry Ring for details. Special events will need pre-booking.

Yearly event or point of interest Carbon-gain visitor centre with innovative design surrounded by rare limestone grassland and ancient woodlands. Seasonal activities – see website.

Directions **Car:** From the M5 Junction 13 take A419 towards Stonehouse. 1 mile to roundabout, take 3rd turning to Eastington. 2 miles to mini-roundabout, 1st turning left towards Frocester. Through Frocester to the top of Frocester Hill. At the crossroads go straight over and turn left after 300 yards towards Nympsfield. After 500 yards you will come to crossroads: go straight over, signposted towards Nailsworth. After 1 mile turn left at the Thistledown sign. If you reach the wind turbine, you are 300 yards too far.

From the M4 Junction 18: Take the A46 north towards Stroud for approx 10 miles. At the junction with the A4135 (Tetbury / Dursley) turn towards Dursley through village of Kingscote. Half a mile out of the village turn right onto the Stroud road. After 2 miles turn right onto the Stroud / Nailsworth road (B4058). After 1 mile bear left to Nympsfield. After 2.5 miles turn right to Nympsfield. After 500 yards you will come to a crossroads: go straight over, signposted towards Nailsworth. After 1 mile turn left at the Thistledown sign. If you reach the wind turbine, you are 300 yards too far.

Description Thistledown is committed to promoting awareness of and respect for agricultural and environmental practice through exploration of the relationship between land and life. It aims to provide a unique and inspiring way of enjoying a secret valley in the Cotswold countryside, blending adventure and sculpture trails with the natural beauty of the area.

YOU CAN SEE

Biofuel • Exchange heat pump – coming • Solar heating – coming • Wind turbine – full size, next to site • Wood stoves • Straw-bale building • Timber-framed building • Renewable insulation • Compost toilet • Rainwater collection • Greywater and brown water collection • Reed bed • Waterless urinals • Hemp / lime / clay walls • Recycling • Composting.

COURSES

See website for details.

SOMERSET

119. Bath Organic Group

    OD SAT AND TUES, PLAY AREA

Lower Common Allotments, Upper Bristol Road, Bath. Tel: 01225 484472. info@bathorganicgroup.org.uk www.bathorganicgroup.org.uk

Open to the general public Sat and Tues, 10am-1pm.

Entry Free. Please book guided tours and school visits (small fee is charged).

Yearly event or point of interest Growing Green – Spring Bank Holiday Monday. Seed Swap – see website for details.

Directions To the north of the Upper Bristol Road on the Lower Common Allotments. The entrance is to the left of the Victoria Park children's playground opposite the Hop Pole Pub. Limited parking inside the gate near the Trading Hut – walk east along the track to the BOG Demonstration Gardens.

Description Bath Organic Group demonstration gardens, featuring all that is good in organic gardening: vegetables, herb and fruit plots, trading hut, demonstration small gardens, composting area, tree bog, polytunnels, community orchard, bees, wildlife and pond area.

YOU CAN SEE

Compost toilet • Vegetable plots.

YOU CAN DO

Volunteering.

120. Bristol Permaculture Group

Tel: 0117 902 7913. bristol_permaculture@yahoo.co.uk

Yearly event or point of interest Annual 12-week part-time permaculture design course from January, plus regular introductory courses. February 'Seedy Sunday' seed swap event.

Directions Bristol Permaculture Group has a range of projects around Bristol – email for details of projects and directions.

Description An informal group of 200 projects where individuals are actively involved in urban sustainability projects including community allotments, orchards, school gardens, plant and seed swaps, waste reuse, composting, tree-planting, festivals and food-growing projects. The email group allows members to keep informed of courses, events and workshops and to share ideas, information, tools and resources. Email to join or book courses.

YOU CAN SEE (BY APPOINTMENT OR ON WORK DAYS)

• Renewable energy • Biofuel.

COURSES

Example course listing:

Introduction to Permaculture Weekend A 2-day introduction to the principles, ethics and practice of permaculture. Talks, workshops, slide shows and practical activities. Exploring an integrated design approach to organic growing, community action, waste minimisation, localisation, energy and water efficiency and much more. Be part of the solution.

Permaculture Design Course 12 weekly evening sessions plus field visits to a range of inspiring sustainable projects around the south-west. This course covers all aspects of permaculture, culminating in the opportunity to design your own garden or on a local community project.

Other workshops BPG also offers a range of occasional practical workshops including green woodwork, fruit tree grafting, wild food walks, hedge-laying, gardening and more.

UK tuition Their courses have led to the formation of a vibrant, active group. If you would want to bring like-minded people together in your area, they can provide tuition for introductory courses around the UK.

Details of all courses are available by email.

121. Carymoor Environmental Centre (Part of Somerset Wildlife Trust)

Dimmer Lane, Castle Cary, Somerset, BA7 7NR.

Tel: 01963 350143, Fax: 01963 350468.

enquiries@carymoor.org.uk www.carymoor.org.uk

Open Not usually open to the general public, but appointments can be made to access the renewable technology. Opening times: first working Monday of the month, plus open days. Please pre-book group visits.

Entry Charge for group visits.

Yearly event or point of interest Open evenings: first working Monday of every month. Carymoor Open Day and Somerset Wildlife Trust Open Days. Rolling course programme.

Directions **Train:** The centre is located two miles from Castle Cary station, off the B3153 – follow signs for the household waste and recycling centre. **Bike**: A marked cycle route passes close to the site.

Description Carymoor Environmental Centre is based on the restored landfill site just outside Castle Cary in Somerset. Its location provides an ideal setting to contrast environmental sustainability with the less sustainable activity of the landfill. Illustrating renewable energy, sustainable development and education, the centre has developed a reputation that exemplifies environmental good practice.

YOU CAN SEE

Photovoltaics • Wind turbine – full size • Straw-bale building • Timber-framed building • Renewable insulation • Compost toilet • Reed bed • Recycling • Composting • Sensory garden • Pond.

YOU CAN DO

Volunteering • Walks and talks.

COURSES

Examples below – see website for more details.

Hedge-Laying Practical and theory-based course inside the classroom and outside on site; learn traditional hedge-laying skills, find out about the tools of the trade and get hands-on experience. For those with little or no experience.

Also **Reed Beds, Straw-Bale Building.**

CHILDREN'S ACTIVITIES

Various events and activities – see website.

122. Centre for Sustainable Energy

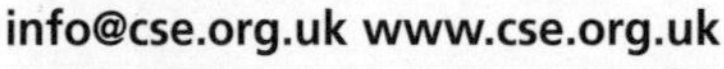

The CREATE Centre, Smeaton Road, Bristol, BS1 6XN.
Tel: 0117 929 9950, Fax: 0117 929 91140.
info@cse.org.uk www.cse.org.uk

Open Not open to the general public.

Directions Located in the historic B-Bond warehouse near to the floating dock in Bristol. **On foot:** The city centre is 1.8 miles away – a pleasant 30-minute walk along the docks. Temple Meads railway station is approximately 2 miles away and Broadmead bus station is just under 2.5 miles away. **Bike**: From the city centre, a cycle track runs alongside Cumberland Road to CREATE. Bicycle racks are provided on site. **Bus:** The Baltic Wharf 500 loop service runs to CREATE via the city centre, Broadmead and Temple Meads at half hourly intervals between 7.40am and 6.10pm. The two stops closest to CREATE are Cumberland Road and across Junction Lock Bridge. **Ferry:** The ferry stops next to the Nova Scotia pub, just a few yards from CREATE. For full details, call the Bristol Ferry Boat Company on 0117 927 3416. **Train**: Temple Meads mainline railway station, serviced by trains from across the country, is situated around 2 miles from CREATE. **Car:** See website for access by car.

Description The Centre for Sustainable Energy is a national charitable company established in 1979 with a board of trustees and around 35 staff based at the CREATE Centre in Bristol. Their mission as an independent charity is to advance sustainable energy policy and practice. They seek energy solutions that engage people and communities to meet real needs for both environmentally sound and affordable energy services. They run courses on sustainable living.

COURSES

Introducing Sustainable Energy A half-day course introducing the basic facts and giving participants the confidence to pass on essential energy advice.

Renewable Energy for Heating Learn how renewable energy sources can be used in domestic heating systems; this 1-day course will provide an overview of three renewable technologies and some basic skills for assessing their feasibility; the comparative running costs of wet central heating systems; solar hot water systems; ground source heat pumps; biomass stoves and boilers.

Renewable Energy Essentials A one or 2-day course in household and community-scale renewable energy systems, including solar water heating, biomass heating, ground source heat pumps, solar photovoltaics and small-scale wind.

Transport Essentials An 1-day course providing the facts, figures and evidence to assess options and/or advise others on local transport and community-scale transport solutions.

123. CREATE Ecohome

OD    OD MON-FRI, RESOURCE ROOM, VIDEO

The CREATE Centre, Smeaton Road, Cumberland Basin, Bristol, BS1 6XN.
Tel: 0117 925 0505, Fax: 0117 922 4444.
create@bristol-city.gov.uk www.bristol-city.gov.uk/create

Open Mon-Fri plus certain weekends for special events (see website for latest information). 12noon-3pm, plus certain evenings for special events. Please pre-book school group visits.

Yearly event or point of interest CREATE and the Ecohome hold an annual Eco-Living fair (May) and many other events and exhibitions throughout the year.

Directions See directions for the Centre for Sustainable Energy above.

Description The CREATE Ecohome is a purpose-built demonstration house designed to inspire and inform people about sustainable living and building. It has been built and furnished using natural, recycled and locally sourced materials, and has many energy-saving features. It also includes a resource room and exhibition area. The Ecohome and adjacent CREATE Centre (which is home to a number of environmental organisations) are owned and managed by Bristol City Council.

Accessibility The Ecohome is wheelchair-accessible on the ground floor, with a stairlift to the upper floor. We have an audio guide suitable for visually impaired visitors.

YOU CAN SEE

Biomass pellet production (coming soon) • Thermal mass • Specialist ventilation • Solar heating • Passive solar design • Wind turbine – domestic (to be installed on adjacent CREATE building) • Timber-clad building • Renewable insulation • Rainwater collection • Greywater system.

YOU CAN DO

Volunteering.

124. Genesis

MEETING FACILITIES

Somerset College of Arts and Technology, Wellington Rd, Taunton, TA1 5AX.
Tel: 01823 252934, Fax: 01823 366741.
genesis@somerset.ac.uk www.genesisproject.com

Open to the public on open days. Can also book technical tours on non-open days at a cost.

Yearly event or point of interest Continuous Professional Development, educational visits. Sustainable focused events throughout the year (check website).

Directions **Coach** and **Bus:** National Express has a regular service from most parts of the country. Berry's is a local service with a wide network of coaches and other buses travelling across the South-West and from London. Buses 22, 22a, 92 and 92a go directly to the College from Taunton bus station (bay 3).

Train: Taunton has a mainline station which is 15/20 minutes walk from the College. There are fast trains direct from London, Bristol, Bath, Birmingham and Cornwall. To catch a bus from the station to the College turn left out of the station onto Station Road. The number 3 bus runs directly to the College. Buses 1, 1a, 2, 3, 25a, 28, 28a and 29 will take you to the bus station, from there buses 22, 22a, 92 and 92a go directly to the College (bay 3). Alternatively take a taxi from the rank outside the train station, the 5-10 minute journey costs approximately £5.

Car: Taunton is just off the M5 motorway and not far from the A303. Buses leave the Park and Ride every 6–12 minutes. The drop-off stop is approximately 100m from the College entrance. Cost: £1.50 return, free parking.

Sustainable transport policy We are committed to promoting awareness of our environmental impact. Our policy includes promoting bus travel, strict car parking management, and improving facilities for cyclists and pedestrians. We recommend that you use public transport to get to Genesis.

Description The Genesis Centre is a new building designed to demonstrate various approaches to sustainable construction. It provides a platform for continuously evolving education and training materials, designed to respond to the diverse needs of the construction industry and raise awareness in the wider community. A conference, training and educational facility for all.

YOU CAN SEE (BY APPOINTMENT)

Biomass production – woodchip • Biomass boiler • Thermal mass • Specialist ventilation • Photovoltaics • Solar heating • Straw-bale building • Cob building • Rammed-earth building • Mud-block building • Timber-framed building • Renewable insulation • Rainwater collection • Reed bed.

COURSES

Examples of short courses running at the Centre include:

Achieving Sustainable Buildings – Clients in Control • Construction and Sustainable Communities • Designing Sustainable Buildings • Engineering Sustainable Buildings • Resource Efficient Construction • Solar Gain and Micro Renewables • Specification and Use of Natural Materials • Strategies for Low-Carbon Construction • Sustainable Water Management in Construction.

125. Musgrove Willows

    OD MONDAY-SATURDAY

Willowfields Lakewall, Westonzoyland, Bridgwater, TA7 OLP.
Tel: 01278 691105, Fax: 01278 699107.
info@musgrovewillows.co.uk www.musgrovewillows.co.uk

Open Mon-Sat, 9.30am-4.30pm, all year round. Site tours by appointment.

Entry Free.

Directions From the M5, take Junction 23 and follow signs for Bridgwater. Westonzoyland is 3 miles from Bridgwater along the A372 towards Langport. In Westonzoyland follow the signs for the Pumping Station Museum and we are ½ mile along Lakewall on the left.

Description A family-run business founded in the 1930s which cultivates and processes over 60 varieties of willow for basket making, soil erosion, lanterns and hurdles. They work sensitively and as organically as possible on their willow beds.

COURSES

Courses are run throughout the year in **beginners' basketry, sculpture, living structures, garden plant climbers** and **hurdle making**.

126. Nettlecombe Court

The Leonard Wills Field Centre, Williton, Taunton, Somerset, TA4 4HT.
Tel: 01984 640320, Fax: 01984 641236.

enquiries.nc@field-studies-council.org www.field-studies-council.org

Open Not open to the general public.

Accommodation 97 beds.

Directions 15 miles from Taunton station. Transport easily arranged to and from Taunton rail and coach stations. 3 miles from A39 at Williton.

Description See 'Field Studies Council' in 'National Organisations' at the back of the book. Eco Centre Status

YOU CAN SEE

Weather station.

COURSES

Arboriculture and Bats: a Guide for Practitioners This course, developed by the Bat Conservation Trust, Arboricultural Association and Lantra Awards, is aimed at arborists, and provides sufficient knowledge to allow them to carry out tree works while considering the potential effects on bats and their habitats.

Bats and Bat Surveys: a Foundation Course for Environmental Consultants Changes in environmental legislation, survey licensing arrangements and PPS9 have all increased the demand for bat surveys and the need for consultants with knowledge about bat survey work. To ensure this demand can be met effectively, the Bat Conservation Trust conducts these highly developed courses specifically for environmental consultants.

Birds and Butterflies of High Summer Birding around Nettlecombe Court is highly productive. Early morning and evening walks are always rewarding, and may even lead to chance encounters with badgers or fallow deer. Within a short drive are the Quantock Hills and Exmoor, offering the chance to see a fine selection of moorland birds from merlins to Dartford warblers.

Family Exploration – Walks, Wildlife and Activities in West Somerset and Exmoor An active week exploring rivers, woodlands, rocky shores and moorland in West Somerset and Exmoor. The evenings provide further opportunities to get up close and personal with wildlife, and to discover your adventurous side.

Identification of Grasses The area around Nettlecombe Court offers a good variety of habitats for looking for different species of grasses: acid heathland, woodland, neutral grassland, a small area of basic grassland and coastal habitats. Through our excursions we will learn which species are associated with which habitats.

Introduction to Moths Explore a range of habitats in Exmoor National Park including deciduous and conifer woodland, grassland, heath, salt marsh and coast. A variety of moth trapping techniques such as dusking, sugaring, larval searches, sheeting and different light traps will be used to capture our specimens. All the moths will be treated carefully and released at an appropriate time after examination.

Trees This course teaches tree identification skills in a particularly diverse setting. Nettlecombe Court has its own arboretum, which provides opportunities to study native species as well as trees introduced from around the globe. Together with visits around the Court and the local Site of Special Scientific Interest (SSSI) there will be talks with slides, group work and question and answer sessions.

There is a range of **other courses** on the website.

WILTSHIRE

127. Heelis

Heelis, Kemble Drive, Swindon, SN2 2NA. Tel: 01793 817400 Fax: 01793 817401.

Open Not open to the general public.

Directions Please phone for directions.

Description National Trust Head Office. See 'National Trust' in 'National Organisations' at back of book.

YOU CAN SEE (BY APPOINTMENT)

New sustainable building • Photovoltaic solar energy system • Natural ventilation • Lime mortar • Thermal mass • Carpet made from wool of National Trust sheep.

YOU CAN DO

Tours – please pre-book.

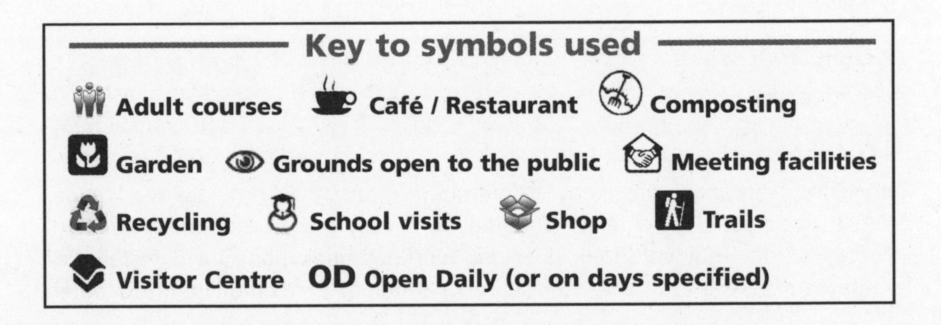

SCOTLAND

FIFE

128. Sustainable Communities Initiatives

      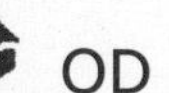 OD

Kinghorn Loch, Kinghorn, Fife, KY3 9YG. Tel: 01592 891884.
inquiry@sci-scotland.org.uk (general enquiries)
tours@sci-scotland.org.uk (to request a tour or book the earthship)
www.sci-scotland.org.uk

Open 7 days a week, 9.45am-4.45pm (10am-5pm Sun). Closed on Mondays throughout the winter.

Yearly event or point of interest Throughout the year SCI runs educational experiences for adults and older children interested in: green living, sustainable homes, living more lightly on the earth. There is a tour of the Earthship and the renewable energy installations, an organic lunch, interactive personal footprinting and also an opportunity to get advice and support.

Description Sustainable Communities Initiatives constructed the first Earthship in the UK. Begun in July 2002, building was completed in July 2004 with the help of over 200 volunteers. An Earthship is a passive solar building with thermal mass, which means that it maintains a comfortable living temperature year round. It is made from natural and recycled materials – predominantly earth-rammed tyres and aluminium cans – and powered by renewable energy such as wind, water and solar power. It catches its own water supply from rainwater, and treats and contains its own sewage in planter beds. Earthships are a concept, not a set design. They can be adapted for any climate worldwide. Earthships offer people the opportunity to build their own homes and make a conscious decision to live lightly on the earth.

YOU CAN SEE

Renewable energy.

YOU CAN DO

Volunteering.

COURSES

Hands-on Learning – Sustainable Communities Initiatives provides work experience opportunities to learn building skills required in Earthship building. These include 1-day volunteer days and working weekends. Our core team of volunteers were part of the building programme of Earthship Fife and worked with Michael Reynolds on the first UK Earthship. For more information on hands-on learning opportunities, join SCI and receive regular updates. SCI aims to set up an accredited training programme as they develop their skill base and future projects. See website for developments.

Creative Waste Workshops Creative Waste Workshops are community-orientated workshops for all ages and families. Workshops are tailor-made for schools, clubs, indoors and outdoors, parties and groups and provide the opportunity to turn household waste into crafts, toys and arts items, using paper, textiles and plastics. We rag-rug, weave, make masks and puppets. Plastic bottles become flowers, windmills – not to mention greenhouses, windbreaks, bus stops and pop stars! We attend festivals, community centres, fun days and schools. The workshops cater for the 5s to over 50s.

Micro Hydro Power Introductory Course 1-day workshop – get a tour of the Earthship and its renewable energy systems, learn about micro hydro technology, explore energy use, learn about grants, suppliers and DIY.

HIGHLANDS

129. Rubha Phoil Forest Garden

Armadale Pier, Isle of Skye, IV45 8RS. Tel: 01471 844700.

sandyru@tiscali.co.uk

Open 7 days a week, dawn until dusk.

Entry £1. Please pre-book guided tours. (No tours available for renewable energy.)

Accommodation Camping.

Yearly event or point of interest Skye music festivals (traditional and pop).

Directions **Train** to Mallaig, Scotland, Calmac **ferry** to Armadale pier, and you're there!

Description Permaculture centre and wildlife sanctuary on 15-acre peninsula with meditative woodland walk.

YOU CAN SEE

Biomass production – woodchip • Specialist ventilation • Solar heating • Passive solar design • Wind turbine – domestic • Timber-framed building • Compost toilet • Rainwater collection • Greywater system • Reed bed.

LOTHIAN

130. Four Winds Inspiration Centre

The Pavilion, Inverleith Park, Arboretum Place, Edinburgh, EH3 5NY.
Tel: 0131 332 2229, Fax: 0131 332 2229. info@four-winds.org.uk
www.four-winds.org.uk www.edinburghtreefest.org.uk

Open Not open to the general public. All access must be pre-booked. Tours and courses need to be pre-booked.

Entry Free to Inverleith Park, 11am-5pm, Sat and Sun. Visits by arrangement.

Yearly event or point of interest Edinburgh Treefest and Woodmarket, 2nd weekend in June – fantastic family event. Crafts, demonstrations, activities, entertainment and much more.

Directions Please contact for directions. Edinburgh Treefest: Inverleith Park, Arboretum Place, North of Edinburgh, adjacent to Royal Botanic Gardens, Stockbridge.

Description Environmental education and craft-based charity offering a wide range of crafts, herbal and renewable energy classes. We also produce individually designed, chainsaw-carved sculptures and furniture using wood from local Dutch Elm disease fellings. Organise Edinburgh Treefest.

YOU CAN SEE (BY APPOINTMENT)

Biofuel • Photovoltaics • Solar heating • Wind turbine – domestic • Interesting site and garden.

YOU CAN DO

Volunteering.

COURSES

Subjects include wood and stone carving, basket weaving, wild wood furniture, felt making, mosaic, and parent and child classes; wind turbine and solar water heating system construction, and making biodiesel. Examples of courses running at the Centre are:

Creams and Ointments Learn how to prepare your own cream or ointment using locally sourced ingredients. Students identify and discuss the plants, habitats, properties, uses etc. and are then guided through the process involved. Also **Parent and Child Craft Classes** and **Herbal classes**.

DIY Solar Panel Heating Make your own solar water heating system over two days: course includes materials, background information on solar water heating, advice on installation and operation techniques for your system, with due attention to safety. You do not need special skills, just the ability to handle the usual domestic tools. Take home a good quality solar water heating collector, which would normally cost 3 times the price if professionally installed.

Felt Making We will show you how to produce felt using natural coloured fleece direct from the sheep. Learn the techniques involved in carding, preparation and felting, to design and produce your own hat or wall hanging.

Garden Sculptures and Plant Supports Learn to make sturdy and useful frames and garden sculptures for plants and climbers in your home and garden. Using weaving techniques with willow, cane and hedgerow materials, you will be able to complete at least one article in the day.

Making Tinctures Learn how to make your own herbal tinctures. We will discuss the recognition, collection and preparation of the plants along with their properties and uses, whilst providing step-by-step guidance on how to produce medicinal tinctures. Bring along half a litre of vodka and some jam jars.

Paper Making Enjoy a day making experimental pieces of handmade paper using recycled material and basic kitchen equipment. You will learn the fundamentals of papermaking, investigating colour, texture and smell. It's messy, fun and inspirational.

Wild Wood Weekend This 2-day course combines green woodworking and rustic-furniture making and includes: the basics of turning green wood and rustic-furniture making; designing and completing a project (e.g. a small coffee table or garden seat); choice of wood, preparation, finish etc, along with methods, safe and efficient tool use.

Willow Weaving Using both traditional and contemporary weaving techniques, working with willow and hedgerow materials, learn to weave a wide range of products, e.g. fruit baskets, log and kindling baskets, hedgerow baskets, random weave baskets and spheres, willow bird houses and feeders.

Wind Turbine Construction (6 days) Learn about small wind turbine design and safe construction. We would hope to complete one wind turbine during the project, and make progress with others. Course includes both theory and hands-on experience: carving wooden blades; winding coils and fitting magnets into purpose-built alternators; wiring; fabrication; erection, etc., as time allows. The finished turbine will be available for a course participant to take home – subject to group vote / material costs covered.

Women's Herbal Learn how we can help ourselves in the changes of womanhood. The first part of the day will involve tasting herbs and exploring the importance of trusting our intuition, including a discussion on the effect on our health of what we eat. Explore the herbs in the Four Winds garden and gain understanding of their role in supporting our process through the cycles.

Wood and Stone Carving Covering the fundamentals of carving, including choice of wood / stone, tools, safety procedures, design, finish. Guided carving will take place throughout the day along with problem solving.

CHILDREN'S ACTIVITIES

Parent and Child Crafts and Herbal Classes We run a full range of crafts and herbal classes for parents and children – e.g. felt making, willow work, paper making, raku ceramic firing, mosaic, exploring the world of herbs.

131. Water of Leith Conservation Trust

 OD

Water of Leith Visitor Centre, 24 Lanark Road, Edinburgh, EH14 1TQ.

Tel: 0131 455 7367, Fax: 0131 443 1682.

admin@waterofleith.org.uk www.waterofleith.org.uk

Open 7 days a week, 10am-4pm.

Entry Free (donations welcome). Groups should pre-book.

Yearly event or point of interest 12 miles of river walkway to explore. There is a programme of river-themed events for children and adults throughout the year, including an annual river clean-up day and Plant Sale.

Directions 3 miles west of Edinburgh City Centre on Lanark Road A70. We are in Slateford area, opposite the Tickled Trout Pub. **Bus** 44, 34 and 20. 300m west of Slateford train station.

Description Working to conserve and enhance the Water of Leith as a haven for biodiversity and an educational and recreational resource for all. The first river charity to be established in Scotland, guardians of the river working to raise the profile of this key environmental asset and promote community action to help the river. The interactive Visitor Centre, the Water of Leith Walkway, and extensive events, education and group visit programmes enable everyone to discover more about Edinburgh's hidden natural asset. (Please call for details of formal and informal education programmes.)

YOU CAN DO

Volunteering – regular volunteer clean-ups and habitat creation projects to improve the health of the river valley.

COURSES

Wildlife and Wild Places This programme covers bugs, slugs and beasties, rivers, water and woodland places. This includes a visit to the Interactive Exhibition, activity walk and drama or art workshop.

Interpreting the Environment through Art This involves visual art workshops with a choice of: observation and recording in the environment; visual art activities using a variety of skills and media; designing, experimenting and making of puppets; light and shadow special effects; mobiles and models and natural sculptures.

People, Place, Past and Present This includes canals and transport, Old Edinburgh, Jacobites in Edinburgh, New Town Georgians, Victorian Edinburgh, and ships, shores and seafarers of Leith. Opportunity to take part in a drama or art workshop and a trail. Some activities also include a slide talk.

MORAYSHIRE

132. Findhorn Foundation

OD, ART STUDIOS

The Park, Findhorn, Forres, Morayshire, IV36 3TZ.
Tel: 01309 690311, Fax: 01309 691301.
enquiries@findhorn.org www.findhorn.org

Open 7 days a week. Site always open. Tours of the site throughout the summer every day except Tues and Thurs at 2pm. Tours of the Living Machine (on-site waste water treatment facility) every Tues and Thurs at 2pm.

Entry Free (tours are £2). Please book group visits.

Accommodation B&B, camping.

Yearly event or point of interest The Foundation runs a year-long programme of workshops, as well as tours of the site throughout the summer.

Directions From Forres (on the A96 between Inverness and Aberdeen), follow the signs towards Findhorn. The Foundation is one mile before the village of Findhorn.

Description The Findhorn Foundation and Community is home to more than 400 people who are exploring new ways of living. It is a spiritual community, a centre for holistic learning, and a developing ecovillage. Educational programmes offer practical steps for personal and planetary transformation, while the ecovillage is a demonstration centre for new directions for humanity and the planet.

YOU CAN SEE

Biomass boiler • Biofuel • Exchange heat pump • Photovoltaics • Solar heating • Passive solar design • Wind turbine – full size • Earthship • Straw-bale building • Timber-framed building • Renewable insulation • Greywater system • Reed bed • Universal Hall (performing arts centre) • Pottery • Art studios.

COURSES

A year-round programme of workshops reflect the core values of connecting with the sacred, both within ourselves and throughout the natural world. They are a chance for personal growth, to meet others who want to live consciously and creatively, an invitation to let go of your limits, open to love and to 'be the change' you want to see in the world.

Ecovillage Training Yearly month-long training designed for those who are beginning to explore sustainability as well as those actively engaged in existing projects.

Experience Week Starting most Saturdays, this week is an introduction to our diverse and vibrant community. It is an opportunity to take part in many aspects of everyday life, and to experience the spiritual principles on which the community is founded.

Exploring Community Life Serving many purposes, this week allows you to stay on after another programme, return here to reconnect, recharge your batteries, or get a deeper sense of what it is like to live and work here in preparation for a longer stay

Life Purpose This week is intended to help you understand your life's purpose – pointing the way for you to know who you truly are and what makes your life truly meaningful.

Spiritual Practice This week is designed to give time to explore ways which can help us connect with spirit in its many aspects, support us in our daily lives, and guide our paths of growth and transformation.

For further details on these, or our other workshops and special events, visit our website www.findhorn.org, or phone 01309 690311.

PERTHSHIRE

133. Kindrogan Field Studies Centre

  4 WORKROOMS, LIBRARY

Enochdhu, Blairgowrie, Perthshire, PH10 7PG.

Tel: 01250 870150, Fax: 01250 881433.

enquiries.kd@field-studies-council.org www.field-studies-council.org

Open Not open to the general public.

Accommodation 95 beds.

Directions **Train**: Direct train from London Kings Cross (leaving 12 noon) to Pitlochry (arrive 6.30pm) on Inverness line • Minibus transfer to Centre from Pitlochry 10 miles.

Description See 'Field Studies Centre' in 'National Organisations' at the back of the book.

COURSES

A Weekend of Bats This weekend introduces all the British bats, with the emphasis on species found in Scotland. Ecology, breeding, feeding, hibernation and conservation will be covered during the course. Bats will be investigated from a practical angle, with the theory and use of ultrasonic bat detectors in the country around Kindrogan where at least four bat species can be found. The course will include the life cycles of British bats, habitats, identification theory and practice from specimens, photos, visits to local roosts, bat sounds and live bats on the wing.

Birdcraft This birdwatching course improves your chances of making a correct identification and helps you to appreciate the bird and its habits. It will teach you to look at the whole bird and not just its colours and patterns, and how the use of fieldcraft improves your view of birds.

Fern Identification Kindrogan is well placed for finding a wide variety of ferns and fern allies, with several uncommon species in the neighbourhood. Advice is given on suitable books, the use of keys, and which ferns are most likely to be found in particular habitats. Species are viewed in context, and other plant groups, birds, land use and the underlying geology are also discussed.

Hunt the Haggis! – Family Environmental Eco Adventure Follow the haggis trail on a trip to the highlands; experience the wilderness and watch a bit of wildlife along the way. Take part in challenges, meet new friends and discover if the legend of the haggis is still true.

Introduction to Mosses Kindrogan has a good variety of habitats within walking distance, and a wide range of bryophyte (moss and liverwort) species. This weekend will look at: the identification of common bryophytes in the field; the variety of habitats in which bryophytes grow; and the use of microscopes and stereoscopes to further one's knowledge.

Lichen Identification This lichen course is aimed at beginners, those who have a little knowledge and those who need to know as part of their work (conservationists, rangers, wardens, site managers). The course covers some general aspects of the ecology of lichens, site and habitat management, and will help you gain confidence with using keys and microscope work.

NVC: Woodlands On each of these days there are visits to local sites with a range of different woodland types. Find out how to classify their vegetation using the National Vegetation Classification (NVC) and in doing so consider the ecology of the various woodland types, and their place in a wider British context. As well as the field visits there are indoor sessions in the evenings.

Spring Birds Learn the sounds of native Scottish birds as well as identifying them. Visit several nearby sites including Killiecrankie, Loch of the Lowes and Loch of Kinnordy, and spend some time around the Centre itself, where we will have one 'Dawn Chorus' morning when the birdsong will be at its best.

Walk on the Wild Side Want to learn navigation? Want to summit a mountain? Learn survival skills and how to build a shelter in the local woodland, plan your expedition and route then set out into the wilderness with your possessions on your back. Set up camp and enjoy hot chocolate after bagging a peak, share your experiences and look forward to a well-earned sleep!

Further **species identification** courses – see website.

WALES

BRIDGEND

134. Welsh Biofuels Ltd

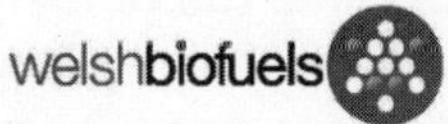

32 Chilcott Avenue, Brynmenyn Industrial Estate, Brynmenyn, CF32 9RQ. Tel: 01656 729714 Fax: 01656 728178
ken@welsh-biofuels.co.uk www.welsh-biofuels.co.uk

Open Not open to the general public – all visits must be pre-arranged.

Directions Available when booking.

Description The premier UK manufacturer of wood pellets to be used in biomass heating systems, offering a viable and sustainable alternative to the rapidly diminishing supply of fossil fuels. Working with clients in the corporate, industrial and domestic markets, Welsh Biofuels Ltd delivers biomass solutions that are transforming the heating sector throughout the UK.

YOU CAN SEE (BY APPOINTMENT)

Renewable technology and building design • Biomass production – pellet • Biomass boiler.

CAMARTHENSHIRE

135. Designed Visions

42 Llandeilo Road, Brynamman, SA18 1BG. Tel: 01269 822786.
enquiries@designedvisions.com www.designedvisions.com

Yearly event or point of interest The website is the main focal point for news of all our courses, services and events.

Directions Activities are held at various locations – contact for information.

Description A cooperative of experienced permaculture teachers and designers, offering courses for adults, design and consultancy. The creatively-taught courses cater for all levels of experience, from total beginners to experienced designers. A collective experience also means that they can offer permaculture consultancy and design services.

COURSES

Some courses are run at residential venues and others as day / weekend events.

Courses that we currently run:

Introduction to Permaculture (1 or 2 days) This course provides an understanding of the ethics and some of the principles underlying this ecological design methodology, along with an introduction to the design process.

Full Permaculture Design Course (90 hours as a 2-week residential or weekends) This course provides a full grounding in permaculture theory, a detailed look at important topics such as soil, and a valuable opportunity for some supported design practice. We also visit established permaculture sites during each course to see established designs in action.

Advanced Permaculture Design Course (4 or 5 days) An opportunity to revise and build on the skills gained during your full design course, and also learn some new techniques.

Living with the Land (4 days – spring, summer and autumn modules) An opportunity to learn how to take permaculture design ideas and to develop a successful land-based project / business. This practical course covers all the aspects of setting up and maintaining a land-based project and also provides a chance to live the lifestyle for a few days.

Training of Teachers Course (4 or 5 days) This course is for anyone wanting to to share their permaculture knowledge with others. We cover course convening, timetable structuring, session planning, creative teaching techniques, and much more.

CONWY

136. Dinas offices, Betws-y-Coed

Dinas, Betws-y-Coed, LL24 0HF. Tel: 01690 713300.

Open Please contact for visit.

YOU CAN SEE

Wood energy scheme.

137. Rhyd-y-creuau, The Drapers' Field Centre

     4 WORKROOMS

Betws-y-Coed, Conwy, LL24 0HB. Tel: 01690 710494, Fax: 01690 710458.

enquiries.rc@field-studies-council.org www.field-studies-council.org

Open Not open to the general public.

Accommodation 90 beds.

Directions **Train:** 1 mile from Betws-y-Coed station. **Car:** 1 mile from A5/A470 junction at Betws-y-Coed.

Description Experience a sustainable lifestyle in the Eco-Centre, a Georgian stone house in two hectares of wooded grounds in the Conwy Valley. For more information see 'Field Studies Council' in 'National Organisations' at the back of the book.

YOU CAN SEE

Weather station.

COURSES

Beginning Bushcraft This course teaches some of the fundamental skills needed to survive in the outdoors. The emphasis is on learning skills and understanding the natural environment, rather than on endurance or hard core survival! Course includes: using bushcraft tools, all within a framework of safety; making some useful items; good environmental practice; collecting and purifying water and the principles of fire lighting.

Bushcraft for Families See 'Beginning Bushcraft' above. With the residential option, it is ideal for families and adults alike.

Environmental Awareness for Outdoor Leaders This course is aimed at people who lead, or wish to lead, groups in outdoor adventurous activities. It includes: activities to help interpret and bring the natural world to life for your groups; exploring ways of reducing the environmental impact of your work; an introduction to the Eco-centres Award and demonstration of practical examples of the award in action at Rhyd-y-creuau.

Family Natural World Adventure This holiday course mixes fun, adventure and wildlife discovery in Snowdonia. Discover and learn about some of the wildlife which inhabits the varied environments of North Wales from mountains to coast. It includes: an introduction to rock climbing and hillwalking, paddling your own canoe on lakes hidden deep in the mountains; walking to find spectacular waterfalls and hearing some of the local myths and legends.

Family Wildlife Discovery Splash about in streams, wander through woodlands and poke about in rock pools – all with the aim of discovering the amazing variety of life from bugs to birds in the Welsh countryside. North Wales has a huge variety of habitats to explore – from high mountains to wild coastlines, tumbling upland streams to gentle pools.

Mammal Identification This course teaches you how to identify all of Britain's land mammals – from sightings and field signs (including calls, droppings, feeding remains and footprints). Led by enthusiastic and knowledgeable tutors, it offers lectures, slide shows, evening activities and hands-on experience doing owl pellet analysis and Longworth live-trapping for small mammals.

Rare Plants of North Wales North Wales contains a diverse range of plant communities and a wide range of habitats as a result of its coastal and mountain sites, its complex geology and varied land use. This course takes you to sites containing some of the rarest plant communities and species in the area. Under expert guidance you will discover the botany and ecology behind these fascinating plants.

See website for further courses in **species identification**.

GWYNEDD

138. Craflwyn Centre

Craflwyn, Beddgelert, Caernarfon, Gwynedd, LL55 4NG. Tel: 01766 510120.
keith.jones@nationaltrust.org.uk : www.nantgwynant.org

Open Not open to the general public.

Directions Please car share or use public transport. Please call for details or see website for timetables.

Accommodation The Craflwyn Centre provides a range of accommodation designed for maximum flexibility. Stay in the hall, which was the former Victorian hunting lodge, or the bunkhouse, or a combination of the two.

Yearly Event or Point of Interest Wide variety of events and activities.

Description Craflwyn is a place to get together and to get away from it all. Run by the National Trust (see National Organisations in back of book), Craflwyn offers facilities for a wide variety of events and activities.

Concern for the environment plays a big part in how Craflwyn is run. All staff and visitors are encouraged to be as 'green' as possible whilst they are here, and hopefully when they return home too.

YOU CAN SEE

Ground source heat pump • Solar panels • Sheeps wool insulation • Sewage treatment plant • Water use reduction • Energy use conservation measures.

YOU CAN DO

Walking holidays • Access to Nantgwynant valley.

COURSES

See website.

139. Margam Park Education Centre

    WORKROOM, CLASSROOM, LIBRARY

Margam Park, Port Talbot, SA13 2TJ. Tel: 01639 895636, Fax: 01639 888659.
margam_sustainable_centre@hotmail.com www.field-studies-council.org

Open Not open to the general public.

Accommodation 40 beds.

Directions Please contact for information.

Description See 'Field Studies Council' in 'National Organisations' at back of book. Eco centre status.

YOU CAN SEE

Exhibitions.

COURSES

Arboriculture and Bats: a Guide for Practitioners This course, developed by the Bat Conservation Trust, Arboricultural Association and Lantra Awards, is aimed at arborists, and gives sufficient knowledge to allow them to carry out tree works taking into account the potential effects on bats and their habitats. Course includes: basic bat biology and ecology; bats and the law; potential tree roost recognition; where to go for help; emergency procedures and practical exercises.

Using Bat Detectors and Sound Analysis Tools This course allows participants to build on knowledge gained on the foundation course or in the field, and provides a comprehensive understanding of bat detectors and sound analysis techniques. Participants learn to use their ears to break down sound to allow more effective use of heterodyne detectors, and are taken through the sonogram analysis process, from making recordings in the field to the identification of bat calls from recordings.

140. Neath Port Talbot – Aberdulais Falls – National Trust (see National Organisations)

Aberdulais, nr Neath, Neath and Port Talbot, SA10 8EU.
Tel: 01639 636674, Fax: 01639 645069.
aberdulais@nationaltrust.org.uk www.nationaltrust.org.uk

Open Varies – please see website.

Directions **Bus**: First 158 Swansea–Banwen, 154/8, 161 from Neath; Stagecoach in South Wales X75 Swansea–Merthyr Tydfil. **Bike**: National Cycle Route 47 passes property. Access near B&Q Neath to Neath Canal towpath and Aberdulais Canal Basin. **Car:** on A4109, 3ml NE of Neath. 4 miles from M4 exit 43 at Llandarcy; take A465 signposted Vale of Neath.

Description See 'National Trust' in 'National Organisations' in back of book.

YOU CAN SEE

Hydro-electric plant • Interactive computer • Fish pass • Observation window • Display panels.

COURSES

See website for list of events.

141. Tir Penrhos Isaf

    PLANTS FOR SALE

Tir Penrhos Isaf, Hermon, Dolgellau, LL40 2LL. Tel: 01341 440256
mawddach@gn.apc.org www.konsk.co.uk

Open Not open to the general public, but guided tours available for up to 7 members by appointment to visit or see the renewable technology. All visits and courses need to be pre-booked.

Accommodation B&B, camping.

Directions Emailed or posted after booking course / making appointment for guided tour.

Description A consciously designed family permaculture holding, first established in 1986 and redesigned in 1991. Main features are wilderness regeneration, upland watershed management, carbon sequestration, low-impact, low-carbon lifestyle, integrated employment opportunities, non-violent horse training, small livestock, food growing, and courses in applied permaculture design.

YOU CAN SEE (BY APPOINTMENT)

Thermal mass • Photovoltaics • Wind turbine – domestic • Water power • Coppice wood heat • Timber-framed building • Renewable insulation • Rainwater collection • Greywater system • Turf roof.

COURSES

Permaculture Design and Low-Impact Development Primarily aimed at those who are interested in establishing low impact communities. The course covers environmental, communal and individual aspects, and includes details of current planning policies.

Permaculture Design Courses: Introductory, Full Design and Advanced Permaculture design is the tried and tested strategy for generating sustainable systems at all scales and levels, as individuals and groups. Founded on the observation of nature, permaculture design provides an ethical framework for evaluating our decisions, uses ecological principles as guides to our actions, and a set of design tools to allow us to create low-energy human lifestyles that are inherently efficient, abundant and fun.

Permaculture Design for Horse Keepers We can use the ethics and principles of permaculture design to reduce the environmental impact of horses, designing systems that mimic their natural patterns and lifestyle as far as possible. As well as looking at environmental factors, the course also includes personal needs, involving the community, recognising limits and product creation.

Practical Permaculture Design for Holdings Aimed at those with some land or those thinking of getting some, the course applies the ethics and principles of permaculture to the design of productive holdings covering environmental, community and individual aspects. Supported by information from our own holding.

Wilderness Regeneration and Harvesting: Interacting with Zone 5 Encouraging the regeneration of wilderness can bring enormous benefits to both the environment and community. Our own individual interactions with regenerating systems can also be immensely rewarding. Such interactions model the permaculture ethics and principles, and can provide a wide range of products and raw materials. This is supported by Argel, our regeneration project established in 1986.

PEMBROKESHIRE

142. Brithdir Mawr Community

Brithdir Mawr, Cilgwyn Rd, Newport, Pembrokeshire, SA42 OQJ.
Tel: 01239 820164 www.brithdirmawr.co.uk

Open Not open to the general public. Please pre-book all visits.

Accommodation Bunkhouse.

Yearly event or point of interest Brithdir Mawr Camp, the chance to spend a long weekend having a go, with music and dancing. Regular guided tours.

Directions Full details for **public transport** to Newport on website, bearing in mind this is Pembrokeshire, not Gwent. From Newport; leaving the village in the Cardigan direction, pass the Golden Lion Pub on the left and take the first right to The Candle Workshop. At the next T-junction turn left and continue along this road, up the hill; ¼ mile after the road starts to descend, look out for the hand-carved sign on the left at the top of the drive.

Description Brithdir Mawr Community is a collective of people working towards sustainability who wish to share their aim with others. They take care of the land, recycle and conserve resources, garden and farm organically, and are off the grid for electricity and water. They strive for a life in which their footprints are as light as they can be.

YOU CAN SEE (BY APPOINTMENT)

Photovoltaics • Solar heating • Wind turbine – domestic • Timber-framed building • Renewable insulation • Compost toilet.

COURSES

More details on all courses including dates and prices are on the website. If one of these courses is not listed, please let us know you are interested. We aim to provide high quality teaching to a small number of participants, and are happy to arrange courses for private groups of 4-6.

Basketry Various courses covering many different skills, including round, square and oval stake and whale baskets, celtic frame baskets, hedgerow basketry and willow-bark baskets.

Circle Dancing Learn traditional dances from around the world, often with live music.

Community Living Experience communal life for a weekend, finding out more about alternative energy production, organic food production and the nitty-gritty of sustainable living.

Greenwood Skills Various courses for both the beginner and more experienced maker. Projects from tent pegs and mallets, gate hurdles, bent-wood chairs and cleft-wood gates.

Herbcraft and Natural Beauty Find out how to make effective healing creams, ointments and cosmetics with all natural ingredients.

Seed Saving Learn the art of gathering seeds from the garden; safeguarding heirloom varieties and freeing yourself from reliance on commercial patented seed companies.

Woolcrafts Courses covering many techniques including spinning, felting, knitting, crochet, rugging and weaving. Looking at all the different ways that fleece can be turned into fabric.

Working Horses Courses for beginners to provide the opportunity to familiarise themselves with horse and harness.

143. Dale Fort Field Centre

   LIBRARY, 5 WORKROOMS

Haverfordwest, Pembrokeshire, SA62 3RD.
Tel: 0845 330 7365, Fax: 01646 636554.
enquiries.df@field-studies-council.org www.field-studies-council.org

Open Not open to the general public.

Accommodation 100 beds.

Description See ' Field Studies Council' in ' National Organisations' in the back of the book.

YOU CAN SEE

Weather station.

COURSES

Large amount of **marine studies.**

Bird Survey Techniques Birds are often seen as a good measure of the health of the environment, and the data produced from accurate, long-term surveys are essential to measure population trends. This is a course for the keen amateur ornithologist or wildlife professional who would like to learn some basic bird survey techniques.

Exploring the Seashore This course looks at the ecology of all types of seashore around the centre, including different rocky shores, saltmarsh, estuary and mudflats, and includes a walk around the clifftops to view seabirds when the tide is in – all at a leisurely pace. There are slide shows in the evening.

Fauna of Sediment Shores This 2-day course concentrates on the identification of infauna and epifauna from sediment habitats. It includes visits to various types of sediment shores, from the mobile sandy beaches of the Atlantic coast of Pembrokeshire to the sheltered mudflats of the Milford Haven.

Marine Plankton This course concentrates on the mysterious organisms of the plankton. The course consists of sampling excursions to various locations in the Milford Haven waterway aboard the Centre's high-speed boat, followed by microscopy work back in the laboratory. Participants will become familiar with the amazing plants and animals of the plankton.

CHILDREN'S ACTIVITIES

Bugs, Beasts and Birds An opportunity for families who enjoy wildlife activities: watching mammals or the seabirds of Skomer Island, seine netting for fish and crabs in the local estuary, pond dipping, sweep netting for rare ladybirds and plankton trawling on the Centre powerboat. Weather permitting, as much time will be spent outside as possible, taking full advantage of the spectacular natural landscape and the flora and fauna.

See website for further courses.

144. Orielton Field Centre

 LIBRARY, 5 WORKROOMS

Pembroke, Pembrokeshire, SA71 5EZ. Tel: 0845 330 7372, Fax: 01646 623921. enquiries.or@field-studies-council.org www.field-studies-council.org

Open Not open to the general public.

Accommodation 110 beds.

Directions **Train**: 3 miles from Pembroke station – transport can be arranged. **Car**: 3 miles from A4075 at Pembroke.

Description See 'Field Studies Council' in 'National Organisations' at the back of the book. Eco Centre Status.

YOU CAN SEE

Weather station.

COURSES

Bats and Bat Surveys: a Foundation Course for Environmental Consultants Changes in environmental legislation, survey licensing arrangements and PPS9 have all increased the demand for bat surveys and the need for consultants with knowledge about bat survey work. The Bat Conservation Trust is conducting these highly developed courses specifically for environmental consultants. The courses include tuition on: bat biology and ecology; legislation and policy;

survey methods; assessing value and impact; mitigation and monitoring, and recognising when to recruit a more experienced bat worker.

Herbal Medicine: Traditional and Modern Herbal medicine uses traditional techniques and locally grown plants to combat some of the problems of modern living. This course includes practical demonstrations using plants collected nearby, explanation of their safe use in the home today, and a visit to Picton Castle and its famous herb garden.

Lichens as Environmental Indicators This course is targeted at those with some knowledge of lichens, but includes simple and practical methodology, so beginners are welcome. Orielton is an ideal site for investigating the great range of lichen habitats in Pembrokeshire, including ancient woodlands, moorlands and sea coasts.

Making Home Wildlife Videos This course gives hints, tips and techniques for shooting and editing wildlife videos that you and your friends will enjoy watching. Weather permitting, the course will culminate in a visit to Skomer Island to film seabirds.

Mosses and Liverworts With its impressive range of habitats, Orielton provides an outstanding opportunity to see a wide range of common bryophytes, as well as many that are special to this part of Britain. The course will be of interest to all students, teachers, conservationists and keen amateurs. Learn how to identify these fascinating and ecologically important plants and, at the same time, discover how to use them as ecological indicators.

Plant Power Workshop The joint FSC and SAPs (Science and Plants in Schools) workshop gives new and exciting practical ideas for the delivery of the secondary science curriculum. There are hands-on activities both inside and outside the classroom. Activities include: new ways of teaching photosynthesis at KS4 and AS/A2; extraction of plant enzymes; ecology for beginners at KS4; managing risk assessments for fieldwork; and using fieldwork for AS/A2 coursework.

Spring Birds in Pembrokeshire Visit woodland, wetland and cliffs, learn how to identify birds by sight and sound, and examine their general ecology. A visit to Skomer Island, with its internationally important seabird colonies is planned, weather permitting. (Boat and landing fees are not included.) This course is suitable for beginners and those with some experience of birdwatching.

Traditional Craft Trio An opportunity to try out three different craft activities and take away a finished product from each. Willow weaving, stained glass, and felt making are demonstrated by experienced tutors, who will help you to develop your own finished piece of work.

Woodland Management and Conservation How a wood is managed affects the wildlife within it: what favours one species may eliminate another. We visit a variety of woodland sites to examine their different management regimes, and the effect on woodland flora and fauna. As the ability to identify current species is a prerequisite of management decisions, some time will be devoted to identification of trees, shrubs, woodland plants and other indicator species.

Working with Natural Dyes Dyes made from natural materials (e.g. some roots, bark, leaves and lichens) give a wide range of subtle and vibrant colours. This hands-on workshop teaches the techniques of dye extraction, use of mordants and working with different types of fabric to produce scores of different tones and shades. The course is suitable for both beginners and improvers.

CHILDREN'S ACTIVITIES

Family Wildlife Weekend (age group 5-10) Rock pools, pond-dipping and mammal trapping are just a few of the activities in this fun-filled weekend. A chance to find out more about the world around you in the spectacular Pembrokeshire countryside. Outdoor visits are followed by a variety of games and craft activities.

See website for further courses.

West Wales ECO Centre

Canolfan ECO Gorllewin Cymru

145. West Wales ECO Centre

   OD MON-FRI

Lower St Mary Street, Newport, Pembrokeshire, SA42 0TS.

Tel: 01239 820235, Fax: 01239 820801.

westwales@ecocentre.org.uk www.ecocentre.org.uk

Open Mon-Fri, 9.30am-4.30pm.

Entry Free. Appointments can be made to access the renewable technologies.

Directions **Train:** Newport **Car:** Newport is halfway between Fishguard and Cardigan on the A487. Turn left off the main road as indicated by brown sign.

Description Environmental education charity promoting issues surrounding energy use and sustainability: 25 years promoting energy-saving, renewable energy technologies, sustainable lifestyles and green building. They work with schools, local authorities, householders and community groups.

YOU CAN SEE (BY APPOINTMENT)

Biomass production – pellet • Photovoltaics • Solar heating • Rainwater collection.

POWYS

146. Centre for Alternative Technology

      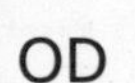 OD

Pantperthog, Machynlleth, Powys, SY20 9AZ.

Tel: 01654 705989, Fax: 702782. info@cat.org.uk www.cat.org.uk

Open 7 days a week. Please pre-book educational groups or special tours and talks.

Accommodation Eco-hostel – accommodation for those on courses.

Directions **By bike or on foot**: Sustrans National Cycle Network – Route 8 passes via the rail station, right up to the entrance of the CAT Visitor Centre. This is a pleasant, albeit a little hilly, route. The journey takes less than half an hour by bike, and a little over an hour on foot. See website for more details.

Train: Machynlleth is on the Birmingham to Aberystwyth train line (Arriva Wales), which can be also accessed via Shrewsbury and Wolverhampton. You can buy an inclusive rail/entry ticket from anywhere in the UK which will give you half-price entry to the Centre. Ask at your local railway station for a MACHCAT ticket.

The Visitor Centre is about three miles north of the Machynlleth train station, so you can continue your journey by bus, taxi, bicycle or on foot. If you arrive by person-powered or public transport you will receive £1 off the normal entry price. For further information about buses and arriving by car see website.

Description Showing practical solutions to environmental problems to carry us into the twenty-first century, the seven-acre Visitor Centre teaches people solutions to environmental problems. CAT also publish books, carries out environmental consultancies, gives free information, and runs courses from 3-day residential courses up to MSc level. CAT incorporates an on-site community, testing the ideas that are preached.

YOU CAN SEE

Biomass production – woodchip • Biomass production – pellet • Biomass boiler • Biofuel • Thermal mass • Specialist ventilation • Exchange heat pump • Photovoltaics • Solar heating • Passive solar design • Wind turbine – domestic • Wind turbine – medium • Wind turbine – full size • Straw-bale building • Cob building • Rammed earth • Mud-block • Timber-framed building • Renewable insulation • Compost toilet • Rainwater collection • Greywater system • Reed bed.

COURSES

Alternative Building Methods • Build Your Own Wind Turbine • Building Integrated Renewable Systems • Building with Straw • Community Renewable Energy Schemes • Convert your Engine to Vegetable Oil •

Domestic Solar Water Heating Systems • Domestic Wind Power Systems • Earth Building • Eco Design and Construction • Educating for Sustainability • Educating with the Eco-Footprint • Energy Awareness • Freshwater Biology • Heat Pumps • Heating with Wood • Heating with Wood Fuel: Installers • Hemp and Lime Conference • Hydro-Electric Power Systems • Install and Test Domestic Photovoltaic Systems • Introduction to Renewables • Make your own Biodiesel • Natural rendering – clay plaster • Organic Gardening • Plant Identification • Renewable Energy Heating Systems • Renewable Energy Systems • Sewage Solutions • Solar Electric Systems • Solar Water Heating for Installers • The Economics of Sustainability • The Sustainable Home • The Whole House • Timber-Frame Self-Build • Water Treatment, Conservation and Recycling • Wind Power • Working with Willow.

Please check the website for further details.

147. Primrose Earth Awareness Trust

Primrose Earth Awareness Trust, Felindre, nr Talgarth, Powys, LD3 0ST.

Tel: 01497 847634. info@primrosetrust.org.uk www.primrosetrust.org.uk

Open Not open to the general public (only on special days).

Directions PEAT is signposted off the A438 just after Three Cocks (when going towards Brecon) and also on the A4078 between Three Cocks and Talgarth.

Description Primrose Earth Awareness Trust is an environmental education charity, working to raise awareness of sustainable food production and other sustainable living practices. They aim to encourage a respect for the earth and an appreciation of the interdependence of people, plants and the environment. Visitors to the Trust often have access to Primrose Farm (a neighbour), which is a nationally recognised exemplar of the application of permaculture principles to fruit and vegetable production and earth care.

YOU CAN SEE

Celtic Roundhouse • Indoor teaching space in a large polytunnel • Circular yurt providing a space for craft activities and storytelling. • Outdoor activity area including willow arbour seating • Eight raised key-hole gardens • Turf-roofed tool-shed • Willow labyrinth forest garden • Successful organic practice growing fruit and vegetables on a small scale.

COURSES

Family Fun Activity Days Activities include maze games, a blindfold trail, making perfumes and potions, art in the yurt, Celtic face painting in the Roundhouse, scavenger hunts, and catching butterflies if you can.

Professional development for teachers Build a stronger team and reduce stress using nature's inspiration. When you are planning your CPD programme, keep this in mind. In this session in the Trust's gardens, your team will work together to learn how to reduce their stress levels through exploring the natural world using all your senses, developing trust within the group, and discovering some gentle meditation and voice exercises. Leave with new ideas for using nature to enhance the emotional well-being of staff and students.

Organic and Sustainable Food Production Theory and practice, demonstrating the processes involved throughout the year in growing healthy food whilst nurturing the earth.

SNOWDONIA

148. Chickenshack Housing Cooperative Limited

Brynllwyn, Rhoslefain, Tywyn, Gwynedd, LL36 9NH.

Tel: 0845 456 5312. permaculture@chickenshack.co.uk www.chickenshack.co.uk

Open Not open to the general public. Please pre-book for tours and visits.

Accommodation Camping; B&B / hotel nearby.

Yearly event or point of interest 2-week permaculture design course in May each year.

Directions Take the coastal road north from Tywyn to the village of Rhoslefain and take the first left in the village, before first house on left. Brynllwyn is exactly half a mile along there, up a short steep road on the right.

Description A fully mutual par value housing co-operative in Snowdonia National Park. The co-operative manages a house, 3 cottages and 5 acres of land to an evolving permaculture design, and is a small community of 6 or 7 people. It is host to an annual permaculture design course and the occasional practical workshop on herbs, yurt making or similar. Please contact in advance if you would like to visit.

YOU CAN SEE

Biomass boiler (30kW) • Solar heating • Passive solar design • Straw-bale outbuilding • Renewable insulation • Compost toilet • Rainwater collection • Natural wetland pond • Willow coppice.

YOU CAN DO

Occasional volunteering opportunities.

COURSES

Study Permaculture Design in Snowdonia A 2-week course exploring sustainability in all its aspects including practical and sustainable design solutions and personal and community development. It takes you from theory into practice; from gardens to communities, on a personal, social and ecological journey. With specialist speakers, site visits to leading projects, practical workshops and much more.

SWANSEA

149. The Environment Centre

OD MON-FRI

The Environment Centre, Pier St, Swansea, SA1 1RY.
Tel: 01792 480200, Fax: 01792 480400.
office@environmentcentre.org.uk www.environmentcentre.org.uk

Open Mon-Fri (and special events), 10am-4pm.

Entry Free. Please pre-book tours of exhibition rooms and meeting rooms.

Yearly event or point of interest Events are held throughout the year along with a rolling programme of exhibitions e.g. Green Christmas Fayre, Sustainable Living Exhibition.

Directions **Train/bus**: 15 minutes walk from Swansea Railway Station, 10 minutes from the Quadrant Bus Station and 5 minutes from the City Centre. If travelling from the east, the Centre can be found off Quay Parade. Turn left just after Sainsbury's towards the Dylan Thomas Centre, turn immediately right in front of Morgan's Hotel, and then left next to the Evening Post building. The Centre is on the left hand side in Pier St. **Car**: If you want to park your car, there is very little street parking in the area. There is a large pay and display car park near the new Sail Bridge: just follow the signs for the Dylan Thomas Centre. There is then pedestrian access to the Centre under the arch through to Pier St.

Description An independent charity focussing on environmental information, education and activity in South Wales, which was set up in 1995. At the Environment Centre you can get inspired, research information in the library or on the internet, buy green goods, or visit our Fairtrade Coffee Shop. Most importantly you can meet like minds and groups.

YOU CAN SEE

Specialist ventilation • Photovoltaics • Solar heating • Passive solar design • Renewable insulation • Rainwater collection • Turf roof on eco-annex.

NORTHERN IRELAND

ANTRIM

150. Ecos Millennium Environmental Centre

      OD, MEETING FACILITIES

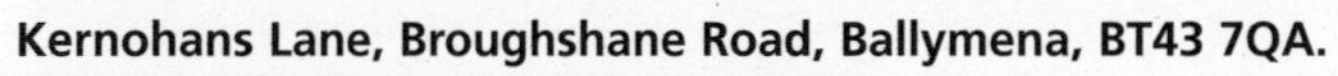

Kernohans Lane, Broughshane Road, Ballymena, BT43 7QA.

Tel: 028 2566 4400, Fax: 028 2563 8984.

info@ecoscentre.com www.ecoscentre.com

Open Mon-Fri, 10am-4pm.

Entry Free. Please pre-book guided group tours and conferencing facilities.

Directions Follow signs to Ballymena and pick up the brown Ecos Centre signage. Junction 11 Ballymena M2 bypass.

Description The Ecos Centre is an educational resource, conferencing venue and visitor centre set in a developing country park where you can explore the natural world and how we use it. Discover what the key issues are, how problems can be tackled, and experience wildlife in its natural habitat. Full corporate hospitality and conferencing facilities available.

YOU CAN SEE

Biomass production – pellet • Biomass boiler • Thermal mass • Specialist ventilation • Photovoltaics • Solar heating • Passive solar design • Wind turbine – medium • Reed bed.

FERMANAGH

151. Derrygonnelly Field Centre

   4 WORKROOMS, DAY CARE FACILITY

Tir Navar, Creamery Street, Derrygonnelly, County Fermanagh, BT93 6HW.

Tel: 028 6864 1673, Fax: 028 6864 1771.

enquiries.dg@field-studies-council.org www.field-studies-council.org

Open Not open to the general public.

Accommodation 65 beds.

Directions **Car**: 13 miles from Enniskillen (B81), 20 miles from Ballyshannon (A46/N3), 115 miles from Dublin, 100 miles from Belfast. **Ferry** Larne – 119 miles (3 hrs). Dun Laoghaire/Dublin Port – 130 miles (3 ½ hrs). Direct links from airports and ferry ports in Britain. Pick-up can be arranged.

Description The Centre is run in partnership with Derrygonnelly and District Community Enterprises, on the site of a former creamery.

YOU CAN SEE

Bird hide • Weather station.

COURSES

Environmental and Heritage Studies This 2-day course is for those who wish to learn more about the ecology, environment and heritage of one of the most beautiful and diverse counties in Ireland. Learn about the vegetational contrasts of limestone and sandstone substrates, blanket bog and Sesleria grasses; discover rare arctic-alpine and boreal plants; encounter blue-eyed grass and see the only Irish occurrence of *Erica Vagens*, a Lusitanian plant; explore everything from potholes and sills to dry valleys and learn about the origin of portal tombs, stone circles, and cup and ring markings.

Fossil Focus Fermanagh has some of the best fossil sites in Northern Ireland. This course introduces the main fossil groups found in Fermanagh, and looks at how fossils can be used to find out what the landscape was like many millions of years ago. The course provides the opportunity to learn more about how fossils can be used and what information can be obtained from them.

Introduction to Trees This residential course deals with the identification of broadleaved and coniferous trees, planting and care of trees, as well as general woodland management including conservation. It is suitable for enthusiastic amateurs, gardeners, farmers, or anyone keen to learn more about trees.

Organic Horticulture This short course provides an introduction to organic horticulture including pests and diseases, fertility, crop rotation, choice of plants and cultivation.

Pollution Monitoring using Stream and River Invertebrates This course introduces you to standard sampling techniques for stream macro-invertebrates, including kick sampling and surber sampling. The macro-invertebrate communities of a polluted and unpolluted stream will be investigated, and the use of biological indices for pollution monitoring will be explained.

Progressive Steps in Wood Turning This course is suitable for those with some experience of wood turning. The topics covered include: the identification of wood types, sourcing and seasoning of wood, and methods of conversion into suitable turning material; safe working practices; demonstration and practice of turning techniques, including bowl and spindle turning and long-hole boring. Finishing and presentation of produce are also demonstrated.

The Carboniferous World The stunning landscape of Fermanagh is home to some of the finest examples of Carboniferous geology in Ireland. The course outlines the global changes that occurred during the Carboniferous period (moving continents, ice age in southern hemisphere, evolution of land plants etc) and how these are recorded in the rocks of Fermanagh.

Trees and Woodlands This course will provide insight into the natural history and ecology of woodlands, and the interactions between commercial forest management and conservation. Identification of woodland plants and animals of semi-natural oak woodland and their interactions will be incorporated.

REPUBLIC OF IRELAND

CLARE

152. Irish Seed Savers Association

 OD

Capparoe, Scarriff, Co. Clare. Tel: 00353 61 921866,
Fax: 00353 61 921397.

info@irishseedsavers.ie www.irishseedsavers.ie

Open Mon-Fri 9am-5pm, Sun 12.30-5pm.

Entry €3 to non-members. Please pre-book tours and school visits.

Yearly event or point of interest Biodiversity day annually, for full list of workshops and courses see website. For gift memberships and sales, please contact.

Directions Drive to the centre of Scarriff, take the Feakle road (left at the roundabout). Drive 2-3 miles along, taking the first right-hand turning marked with a signpost to Seedsavers, which is about a mile down this small road.

Description Membership-based organisation. They aim to: promote the benefits of agricultural biodiversity; to conserve, research and utilise traditional seed varieties; to help keep control of our traditional food resources by growing and distributing Irish-grown seeds and fruit trees; to educate and inform the public on agricultural biodiversity issues both locally and globally; to be a working example of successful organic seed and crop production. Registered charity, non-profit NGO. GM-free, registered organic and bio-dynamic.

YOU CAN SEE

Solar heating • Passive solar design • Timber-framed building • Renewable insulation • Compost toilet.

COURSES

A-Z of Cultivating Vegetables Flow to grow, favourite crops from asparagus to spuds and beetroot to Zucchini.

Beginners' Organic Gardening – Getting Started.

Celt Heritage Day Looking at hedgerow habitats, including coppice crafts traditional skills.

Cob building – 2nd Dimension

Cookery Course – Savoury Selections

Country Wine Making Explore the processes, picking from hedgerows, bottling and sharing recipes of favourites.

Create and Plan for an Edible Sustainable Garden Suitable for any situation: from a rural or small urban garden, school or community garden.

Creating an Orchard The basics of siting, preparing and picking varieties for a suitable orchard site, and tips on the aftercare of fruit trees.

Exploring the Use of Hedgerow Plants

Flora and Fauna Species Identification

Footpath Construction Through native woodlands, and over ditches.

Fungi Day Collecting and eating our native fun guys!

Getting the Best from your Polytunnel

Healthy Cooking with Apples

Hedgerow History Maintenance and ecology.

Hedgerow Maintenance How to plant and care for the traditional hedge.

Heritage week

Introduction to Herbalism with walks to gather and identify them.

Irish Apple Tasting

Native Trees and Shrubs Identification and uses of timber, coppice craft introduction.

Native Woodland Design and Management An introduction.

Open Day at Capparoe Potato talk, seed saving.

Propagating Your Own Plants from Seed Cuttings, clumps and roots, and tips on saving your own seed.

Seed Saving How to!

Starting the Cob Building

Stone Walling Base for a Cob Dwelling An introduction.

Wetlands, Habitat and Species Identification: an introduction.

CORK

153. Green Wood Chairs

The Wooden House, Rossnagoose, Skibbereen, Co. Cork.
Tel: 00353 28 21890, Fax: 00353 28 21897.
Alison@greenwoodchairs.com www.greenwoodchairs.com

Open Not open to the general public.

Directions Details provided when booking.

Description Alison specialises in using 'green' or unseasoned hazel for making chairs, which is coppiced in West Cork. Coppicing is a method of managing woodland whereby the hazel is cut back to the stump and re-grows on a 7-year cycle, producing a truly sustainable material. It is important to her to make chairs in an environmentally friendly way, encouraging the planting and management of more trees and promoting clean technology.

COURSES

Stools and Tables Course (1 day)

Chair making Course (3 days)

Rocking Chair Course (4 days)

Learn to make anything from stools and small tables to a full size rocking chair, using unseasoned, locally coppiced hazel and native hardwoods. No previous woodwork experience necessary; the techniques are very simple. Courses run throughout the year with only 2 students per course, dates are arranged for mutual convenience. All materials and a light lunch and refreshments are included.

154. The Hollies Centre for Practical Sustainability

The Hollies, Castletown, Enniskeane, Co. Cork.

Tel: 00353 23 47001. thehollies@eircom.net www.thehollies online.com

Open Not open to the general public – please book visits and tours.

Accommodation Camping (local B&B).

Directions From Cork by bus (or car) to Enniskeane, 2 miles north on the Macroom road on the right-hand side (wooden gate and polytunnel).

Description The Hollies Centre is dedicated to promoting many aspects of sustainable living and to developing educational programmes for adults and children. A current focus is on natural building (particularly cob), permaculture and woodland management.

YOU CAN SEE (BY APPOINTMENT)

Thermal mass • Solar heating • Tile stove • Passive solar design • Rocket stove • Clay ovens • Rumford Fireplaces • Cob building • Timber-framed building • Renewable insulation • Compost toilet.

COURSES

How to Build Cook Stoves Learn how to build a simple and very efficient 'Lorena' cook stove developed in South America by Ianto Evans, an expert in all questions relating to stoves and fireplaces. The course will also cover the theory of other stoves and fireplaces. Examples of masonry stoves, Rumford Fireplaces and a 'Rocket Stove' can be seen in operation at The Hollies.

Introduction to Cob Building Cob construction is an ancient technique. This hands-on course covers all aspects of cob building – soil testing, cob mixing (by foot and using a digger), wall raising techniques and much more.

Renewable Energy

The Hand-Sculpted House – the Complete Cob Course After this workshop, you will know all you need to be able to go off and build a cob cottage. This 10-day intensive course is hands-on, and you will learn every aspect of the construction of a cob structure: from locating and testing the right kinds of soils for cob, to different ways of mixing it, wall building techniques and all the sculptural details. Get practical experience with windows, doors, arches and niches. Lectures cover siting, passive solar design, drainage, heating, plumbing and electric, natural roofing, and flooring.

DUBLIN

155. Cultivate Sustainable Living & Learning Centre

OD MON-SAT, MEMBER LIBRARY, PUBLIC EXHIBITION GALLERY

15-19 Essex St., Temple Bar, Dublin 8, Ireland.

Tel: 00353 1 674 5773 www.cultivate.ie

Open Mon-Sat, 10am-5.30pm.

Entry Free. Please pre-book tours.

Directions Situated in the West End of Dublin's Temple Bar, between Dublin City Council and Turk's Head Pub on Essex St West.

Description Cultivate is a sustainable living and learning centre dedicated to inspiring healthy, balanced and creative cultural change. The Centre has an eco-shop, information zone, permaculture garden, exhibition gallery, learning lounge, library, and an event hall that is a beautifully renovated church. Visitors can find practical solutions for sustainable living and stay to learn more at our educational events and courses.

Yearly event or point of interest Cultivate's Convergence Festival has grown to be one of the largest and most successful international sustainability events. Convergence takes place around Earth Day (April 22nd) every year.

YOU CAN SEE

Biomass boiler (display only) • Solar heating (display only).

COURSES

Community Powerdown Aims to help communities in energy descent planning.

Convergence Festival Annual gathering includes art, film, conferences, lectures, street fair and more.

Green Building and Energy Seminars and classes exploring renewable energy, eco-renovation and natural building.

Open Source Classes and courses in Linux and other open source software.

Permaculture Introduction and Full Design courses.

Sustainable Economics Course and classes designed to give an understanding of how the global economic system works and the sustainable alternatives.

Workshops Including organic gardening, composting and renewable energy.

LEITRIM

156. The Organic Centre

Project part financed by the European Union
Peace and Reconciliation Programme

OD

The Organic Centre, Rossinver, Co. Leitrim.
Tel: 00353 71 98 54338, Fax: 00353 71 98 54343.
organiccentre@eircom.net www.theorganiccentre.ie

Open 7 days a week 10am-5pm, café weekends only.

Entry €5 / €3. Some things do need pre-booking e.g. tours.

Yearly event or point of interest Hosts Potato Day, Herb Day, Garden Day, a Food Fest and Organic Fair, and an Apple Day, as well as a Green Christmas Craft Fair.

Directions 2 miles from Rossinver on the Kinlough Road.

Description The Organic Centre provides training, education and information on organic growing and sustainable living. The centre was established in 1995 and operates on a 19-acre site on the shores of Lough Melvin. The award-winning ecologically designed Visitor Centre houses training rooms, a shop and the Grass Roof Café. Visitors can wander through 8 acres of gardens including a herb garden, kitchen garden, 8 polytunnels, an orchard and many other attractions.

YOU CAN SEE

Biomass boiler • Passive solar design • Timber-framed building • Renewable insulation • Reed bed.

COURSES

An Introduction to Alternative Energy • Apple Day • Basket Making • Basket Making – Frame Baskets • Building a Garden Pond • Cheese Making • Christmas Decorations with Natural Materials • Cooking and Healing with Wholefood • Cooking with Seaweeds • Cooking with the Grass Roof Café • Cooking with the Grass Roof Café – an Introduction to Cooking • Cooking with the Grass Roof Café – Eastern Cuisine • Cooking with the Grass Roof Café – Winter Vegetables • Discover Wild Herbs – Walk and Workshop • Dry Stone Wall Construction • Dyeing with Natural Colours • Exploring the Links between Blood Group and Diet • Felt Making Weekend • Flowers – their Use and Beauty + Introduction to Companion Planting • Foraging for Free – Wild Food • From Field to Garden – Establishing a Garden from Scratch • Fruit Course – Autumn/Winter Workshop • Fruit Course – Spring Workshop • Fruit course – Summer Workshop • Garden Complete Day 1 • Garden complete Day 2 • Garden Complete Day 3 • Garden Complete Day 4 • Garden Complete Day 5 – Growing Culinary and Medicinal Herbs • Garden Complete Day 6 – Growing Fruit for the Home • Garden Complete Day 7 – Flowers, Companion Planting DIY • Garden Complete Day 8 –

Permaculture, Forest Gardening, Cooking session • Garden Complete Day 9 – Storing and Preserving • Garden Complete Day 10 – Seeds and Crop Planning • Garden Day – Grow Organically • Garden Design for Beginners • Gardening the Heart – Storytelling 1 • Gardening the Heart – Storytelling 2 • Gardening the Heart – Storytelling 3 • Grafting Course • Grains, Breads and Sourdough • Green Christmas Craft Fair • Grow Organic, Cook Organic – Day 1: Winter Vegetables • Grow Organic, Cook Organic – Day 2: Summer Vegetables • Grow Organic, Cook Organic – Day 3: Herbs • Grow Organic, Cook Organic – Day 4 • Grow your own Mushrooms • Growing Herbs – an Introduction • Growing in Polytunnels – a 1 day introduction • Hedge Establishment • Herb Day – Herbs and Health • Herbal Cream Making Workshop • Herbs for Life – Using Herbs Medicinally • Homemade Remedies for the Winter Months • How to Make the Most of 1 acre • Humanure Workshop with Mike Harris • Installation of Solar Electric and Solar Heating Systems • Introduction to Alternative Energy • Introduction to Beekeeping • Introduction to Micro Hydro • Learn How to Live the Good Life – 4th year running • Living Willow Sculpture • Making Green Garden Furniture • Midweek Cooking with the Grass Roof Café • Midweek Cooking with the Grass Roof Café – Eastern Cuisine • Midweek Grass Roof Café Cooking – an introduction to Cooking • Mushroom Hunting • Mushroom Identification day 1-3pm • Natural Cosmetics and Soap Making • Natural Paints – how to make and use them • North-West Food Fest • Organic Gardening for Beginners • Organic Gardening in Schools • Painting with Watercolours • Permaculture course and Permaculture conference • Planting a Small Woodland • Potato Day • Pottery for Beginners – Gardenware • Poultry for the Home • Preserving and Storing Vegetables and Fruit • Reed Bed Systems • Seaweed Extravaganza • Silk Painting Workshop • Slow Food Celebration: Good, Fair, Clean • Spinning for Beginners • Sustainable House Design and Construction • Sustainable Self-Build Workshop • The Complete Polytunnel Course Day 1 • The Complete Polytunnel Course Day 2 • The Complete Polytunnel Course Day 3 • The Organic Fair and Harvest Festival • The Zero Waste Concept – an introduction • Tree Seed Collection • Wholesome Cooking for Children and Teenagers • Wildlife Gardening • Wine Making • Wood Energy – the smart way to heat your home • Wood Energy from Farm Forests.

COURSES IN CO. CLARE

Garden Complete A month-by-month guide running on a monthly basis – check website.

Ground Preparation for the next year (new)

Planting Hedges

Professional/Commercial Training in Market Gardening for Prospective Growers (3rd year running)

Solar, Wind and Water An introduction to alternative energy.

The Complete Polytunnel and Small Glasshouse Course – Clare edition.

Working with Horses (3rd year running)

LIMERICK

157. An tIonad Glas Organic College

Dromcollogher, Co. Limerick. Tel: 00353 63 83604, Fax: 00353 63 83903.
courses@organiccollege.com www.organiccollege.com

Open Not open to the general public without an appointment. Available Mon-Fri 9.30-11.30.

Accommodation B&B.

Yearly event or point of interest Open Day in Spring – see website for details.

Directions The town is 10 miles east of Newcastlewest, 45 miles north of Cork, and 30 miles south of Limerick.

Description The Organic College runs courses in organic horticulture with sustainable living skills to National Certificate level. A further year Diploma course in Organic Enterprise is also run. The College promotes co-operative living skills and at least half of the students' course time is engaged in developing field skills.

YOU CAN SEE (BY APPOINTMENT)

Straw-bale building • Cob building • Mud-block building • Timber-framed building • Renewable insulation.

COURSES

Certificate in Organic Growing and Sustainable Living Skills NVC Level 5. One Year full-time course (part time options available). This course is the only one of its kind in Ireland, and runs from mid September to late May. It features lectures, demonstrations and group discussions, field trips and tours.

Diploma in Organic Enterprise The Diploma course is normally completed over a two-year period. In the first year you complete one of our three certificate courses – Farming, Horticulture or Sustainable Development. The diploma builds on the practical skills developed in these courses and and also features: sustainable farm management; organic crop production and protected cropping; organic enterprise management; work placement and mentoring; countryside management.

Short courses for specialist groups These courses are tailor-made for your needs (e.g. FAS, local authority staff, summer camp groups) and include:

Special Interest Courses

Day and Weekend Workshops

Earth Education Modules

School Tours with Workshops

NATIONAL ORGANISATIONS

Field Studies Council

FSC
Bringing Environmental Understanding To All

Head Office, Montford Bridge, Preston Montford,
Shrewsbury, Shropshire England, SY4 1HW
Tel: 0845 345 4071 (local rate phone call – UK only) or 01743 852100.
enquiries@field-studies-council.org www.field-studies-council.org

Description The Field Studies Council (FSC) is a pioneering educational charity committed to bringing environmental understanding to all.

The FSC believes the more we know about the environment, the more we can appreciate its needs and protect its diversity and beauty for future generations

Established in 1943, the FSC has become internationally respected for its national network of 17 education centres, international outreach training projects, research programmes, information and publication services and wide range of fascinating professional training and leisure courses. They provide informative and enjoyable opportunities for people of all ages and abilities to discover, explore, be inspired by, and understand the natural environment.

The FSC's mission is bringing environmental understanding to all and in working towards this strives to operate in an environmentally responsible and sensible manner.

This commitment to environmental responsibility is demonstrated by the majority of FSC Centres having achieved the Eco-Centre award, which covers all aspects of a Centre's operation. Most Centres do more than simply enough to gain the award. FSC Rhyd-y-creuau, for instance, has conducted a pilot study to becoming carbon neutral and has introduced an innovative cooked waste composter to further reduce waste. FSC Kindrogan was a finalist in a national environmental awareness award and FSC Castle Head is trialling a wind generator and solar panel to run the weather station.

COURSES

By joining an FSC course you will see examples of how we can reduce human impact on the environment in our day-to-day life and they hope that you will put some of this into action when you return home.

During 2007 the FSC will be offering nearly 600 Leisure Learning and Professional Development Courses at 14 of its centres throughout the UK. Full details of all courses are given on the website and in FSC Natural History and FSC Arts courses brochures.

Course categories include:

Crafts and traditional skills • Understanding birds and other animals • Understanding and exploring the natural environment • Flowers and other plants • Habitats and conservation • Families courses – Wildlife and Discovery, Creative, and Active courses • Painting and drawing • History and archaeology • Photography.

Groundwork

Lockside, 5 Scotland St, Birmingham, B1 2RR.
Tel: 0121 236 8565 Fax: 0121 236 7356
info@groundwork.org.uk www.groundwork.org.uk

Description Groundwork is a federation of Trusts in England, Wales and Northern Ireland, each working with their partners to improve the quality of the local environment, the lives of local people and the success of local businesses in areas in need of investment and support. Each Groundwork Trust is a partnership of the public, private and voluntary sectors with its own board of trustees.

Groundwork's vision is of a society made up of sustainable communities which are vibrant, healthy and safe, which respect the local and global environment and where individuals and enterprise prosper.

Groundwork's purpose is to build sustainable communities in areas of need through joint environmental action.

We aim to do this by developing and delivering partnership programmes and projects that deliver benefits equally for:

People: creating opportunities for people to learn new skills and become more active citizens;

Places: delivering environmental improvements that create cleaner, safer, greener neighbourhoods;

Prosperity: helping businesses and individuals fulfil their potential.

Volunteering Groundwork Trusts regularly involve volunteers, especially people seeking work experience before moving into full-time paid employment. Volunteering opportunities will depend on the types of projects being run by individual Trusts. Undertaking voluntary projects can put people in a better position to obtain full-time work – in fact many of Groundwork's existing staff began their careers with us by volunteering.

COURSES

Learning, citizenship and sustainability

We work with children in and out of school and with adults to illustrate how our individual actions can make a difference to our immediate surroundings and the global environment. We help and train teachers to deliver education for sustainable development, for example by involving pupils in improving their school grounds or by establishing links with local businesses. We also educate people of all ages to act more responsibly with regard to energy, waste, water and transport, to conserve natural resources and to respect local places and other people. This includes training businesses in environmental management and giving people the skills to get more involved in local regeneration.

Learn to live differently

Groundwork is one of the main deliverers of an NCFE-accredited foundation level certificate in sustainable development.

Sustainable Development is a term used quite freely today, but what does it mean? How we live, where we shop, what we eat, how we travel, how much electricity we use, what we throw away, what we recycle - it's all important, not just to us as individuals, but also to our wider environment and community. Individual actions and how they contribute to local and global issues, now and in the future, is what sustainable development is all about.

To help raise awareness of sustainable development and enable people to use this awareness to live more sustainably at home, school or in the workplace, Groundwork can offer an accredited training course in sustainable development.

The course involves 30 hours of guided learning and combines group work, team building sessions and individual activities which are designed to inspire, motivate, challenge and be a lot of fun. The course is designed for anyone over 14 and leads to the achievement of the NCFE Foundation Certificate in Sustainable Development.

The National Trust

www.nationaltrust.org.uk

Overview

The National Trust is a charity and is completely independent of Government. They rely for income on membership fees, donations and legacies, and revenue raised from commercial operations.

The Trust has 3.4 million members and 43,000 volunteers. More than 12 million people visit their pay-for-entry properties every year, while an estimated 50 million visit open-air properties.

The Trust protects and opens to the public over 300 historic houses and gardens and 49 industrial monuments and mills.

But it doesn't stop there. They also look after forests, woods, fens, beaches, farmland, downs, moorland, islands, archaeological remains, castles, nature reserves, villages – for ever, for everyone.

The National Trust is committed to setting an example in sustainable practice throughout all its activities, particularly through microgeneration.

The Trust has been using renewable energy for many years, using waterpower for milling and woodfuel for heating. Their first purpose-built hydro-electricity scheme was at Aberdulais Falls, as far back as 1986, and since then over 20 sites have been developed for woodfuel heating, solar, ground heat and hydro-electricity.

Solar energy systems have been installed at six major properties, including Gibson Mill in North Yorkshire, and Craflwyn Castle in Gwynedd. There are eight heating systems using wood rather than oil or gas heating, including Sheringham Park in Norfolk, and the Dudmaston estate offices.

The National Trust's sustainability policies are demonstrated through its central office, Heelis, which has met high quality benchmarks for sustainable design. The building won the RIBA award for sustainable architecture in 2006, given to the building which embodies most elegantly and durably the principles of sustainable design.

The office is cocooned within 1,554 photovoltaic (solar) panels, and combines daylighting with solar capture, generating electricity for the building to run on, with the levels of energy generated displayed in the reception area. All of the timber in the building was sourced from sustainable woodland, much from National Trust properties, and the carpets were specially developed using Herdwick sheep from National Trust land. Heelis generates just 15kg of carbon dioxide per square metre per year, compared to 169kg for a typical air conditioned office.

Places to see the National Trust's renewable energy sources in action include:

Aberdulais Falls Centre, Neath and Port Talbot

Botallack Count House, Cornwall

Bowe Barn offices, Borrowdale, Lake District

Brancaster Millennium Activity Centre, Norfolk

Craflwyn Centre, Gwynedd

Dinas offices, Betws y Coed, Conwy

Dudmaston Estate offices, Shropshire

Gibson Mill, North Yorkshire

Heelis, Swindon

Houghton Mill, Cambridgeshire

Sheringham Park visitor centre, Norfolk

Studland Education Centre, Dorset

Westley Bottom East of England regional office, Suffolk

The National Trust also uses reed-bed systems and waterless urinals in various properties.

COURSES

The National Trust provides fantastic opportunities for getting involved with sustainable projects, as well as for taking advantage of the wealth of knowledge and expertise.

There are courses in anything from basket-weaving to hedge-laying, which are run regionally and can be found on the National Trust website, www.nationaltrust.org.uk.

Three centres run courses all year round; Brancaster Millennium Activity Centre, Chedworth Roman Villa, and Quarry Bank Mill and Styal Estate – these locations offer both adult and family courses, and can be contacted via the website events pages. Brancaster Millennium Activity Centre particularly concentrates on reducing our impact on the natural world, offering opportunities for students to monitor their individual environmental impact.

The Trust also offers careerships for potential gardeners or wardens, these are unique three-year courses blending college work with practical experience at a National Trust property. The careerships culminate in nationally accredited qualifications including NVQs in Amenity Horticulture for Trainee Gardeners or NVQs in Environmental Conservation for Trainee Wardens. There are also vocational training courses available for volunteers and GNVQ students.

The Wildlife Trusts

The Kiln, Waterside, Mather Road, Newark, Nottinghamshire, NG24 1WT.

Phone: 0870 036 7711 Fax: 0870 036 0101.

enquiry@wildlifetrusts.org www.wildlifetrusts.org

Description There are 47 local Wildlife Trusts across the UK, the Isle of Man & Alderney, working for an environment rich in wildlife for everyone. The Trusts manage a total of 2,200 nature reserves, covering more than 80,000 hectares; They stand up for wildlife; they inspire people about the natural world and they foster sustainable living.

Visiting a Wildlife Trust

Many of the Trusts have Visitor Centres where you can enjoy exploring interactive displays or find out more about the local wildlife inhabitants. Go to the main website, where you will find a list of the different Trusts, their websites and where to find them.

Volunteering With 670,000 members, they are the largest UK voluntary organisation dedicated to conserving the full range of the UK's habitats and species. Volunteering alongside some of their members is the easiest way to learn new skills and also to understand better the environment we live in.

EVENTS AND COURSES

Each Trust also has local events and courses, aimed at informing the public about the different local species and habitats, their needs and upkeep. The courses teach new skills for use at home, work or volunteering, which enable individuals to help support wildlife, or to live more sustainably. See individual entries.

GARDENING FOR WILDLIFE

One of the Wildlife Trusts national campaigns is to encourage people to 'Garden for Wildlife'. This will increase habitat for native species and there is information on the website, information packs and many of the Centres have a demonstration Wildlife Garden for visitors to see.

BUILDINGS

Any new centres that the Trusts build are taking the environment into account, reducing energy use as far as possible and using sustainable materials when they are able. This reduces the CO_2 output which causes 'climate change', and also reduces impact on habitats across the world.

WHY CARE?

Wildlife Trusts around the country are concerned about the effect of climate change upon biodiversity on land and at sea. As landowners and managers with considerable local knowledge, they see changes already happening. To learn more about the impacts on our wildlife that could be brought about by climate change go to: www.gloucestershirewildlifetrust.co.uk/index.php?section=campaigns:climatechange

WILDLIFE TRUST SITES OF INTEREST

There are several Trusts which are of special interest, which I have included in the main part of the book.

Bedfordshire: Randalls Farm

Hampshire and Isle of Wight: Curdridge site.

Kent: Romney Marsh Visitor Centre

Nottingham: Attenborough Centre

Surrey Wildlife Trust: Visitor centre

London: Centre for Wildlife Gardening

Norfolk: Cley Marshes

Somerset: Carymoor Environmental Centre

GLOSSARY OF RENEWABLE ENERGY TERMS

Biodiesel

A fuel derived from biological sources such as vegetable oils, which can be used in unmodified diesel cars.

Bioethanol

A mixture of ethanol (produced from grains such as wheat and oil-seed rape) and conventional petrol; normally used to power petrol cars. Ordinary cars can use a blend of 5% bioethanol and 95% petrol, whilst special 'biofuel' cars run on a blend of 15% petrol and 85% bioethanol.

Biofuel

A renewable energy fuel derived from biomass (see below).

Biomass

In energy terms, this refers to the use of a wide variety of organic materials, such as hemp, maize, poplar, willow or sugarcane, for the generation of heat, electricity or motive power.

Heat pumps

These take heat from several metres under the ground (which remains at about 12ºC all year round) and use it to heat a building – just like a refrigerator in reverse.

Geothermal

A heat source from the earth.

Hydrogen fuel cell

Works in a similar way as a battery does, converting oxygen and hydrogen into electricity and water, which it will continue to do as long as these gases are provided to the fuel cell.

Hydro power

Generates electricity from the moving water of a fast-moving stream or river.

Mechanical ventilation heat recovery (MVHR) system

This ventilates buildings by using the stale, moist and warm extracted air to heat the fresh, drier air which is drawn in from the outside.

Passive solar

Uses the heat of the sun to heat a building, usually through glass.

Photovoltaic cells

Convert the energy from sunlight into electrical energy.

Solar heating panels

Use the sun's energy to heat hot water.

Thermal mass

A mass, such as a wall or floor, which absorbs heat during the day, and when the temperature drops, slowly releases the heat into the surrounding area.

Tidal power

Creates electricity like an underwater wind turbine: as the tide flows, the power of the water turns the blades of the turbine.

Wave power

Generates electricity by harnessing the movement of the waves.

Wind turbine

Converts the power of the wind into electricity.

Windmill

Uses wind power to turn cogs to power a device, e.g. a grinding stone or saw.

For more information, see:

www.dti.gov.uk/energy
www.est.org.uk/myhome/generating
www.which.co.uk
www.lowcarbonbuildings.org.uk/home

INDEX OF COURSES

INDEX OF CENTRES AND NATIONAL ORGANISATIONS

YOUR RECOMMENDATIONS FOR THE NEXT EDITION

If you know of a centre that should be included in the next edition of this book, please send us the following information:

Name of centre . . . .
Website (if known) . . . .
Phone number (if known) . . . .
Nearest town . . . .
Any other information . . . .
. . . .
Your name . . . .
Your contact phone number . . . .

Send to: Eco-Centres & Courses, c/o Green Books, Foxhole, Dartington, Totnes, Devon TQ9 6EB.

Also available from Green Books

Water: use less – save more

100 water-saving tips for the home
John Clift and Amanda Cuthbert
80pp in full colour £3.95 pb

Did you know that • we use 70% more water today than we did 40 years ago? • about 95% of water that gets delivered to our houses goes down the drain? This book lists 100 ways in which to do your bit. Tips range from simple measures—such as turning off the tap while you clean your teeth—to more drastic ones, such as installing a rainwater harvesting system. Interspersed with 'Did you know?' facts and photographs throughout, this pocket guide will transform the way you use water.

Energy: use less – save more

100 energy saving tips for the home
Jon Clift and Amanda Cuthbert
80pp in full colour £4.95 pb

We're all using more and more energy: charging up our mobiles and laptops, keeping our rooms so hot that we walk around in short sleeves in the winter, or leaving lights on all day and night. But using so much so freely is causing our climate to change, and our energy bills to rise. This book gives you over 100 tips for saving energy, money and the environment, ideas for taking it a bit further with solar installation or wind turbines, plus facts about our energy use and its impact on our climate.

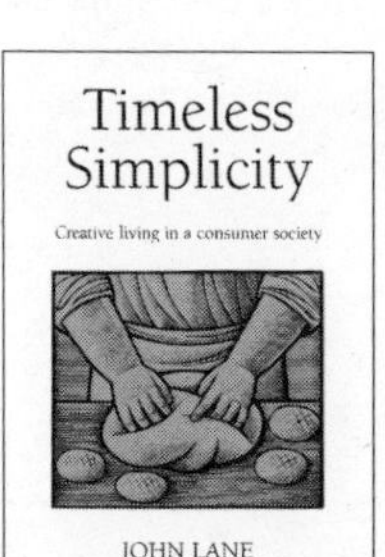

Timeless Simplicity

Creative Living in a Consumer Society
John Lane
Green Books 112pp, illustrated by Clifford Harper
£8.95 pb

"This is a real gem of a book"—Duane Elgin, author of *Voluntary Simplicity*

This is a book about simplicity—about the advantages of living a less cluttered, stressful life than that which has become the norm in the overcrowded and manic-paced consuming nations. It is about having less and enjoying more—enjoying time to do the work you love, enjoying time to spend with your family, enjoying time to pursue creative projects, enjoying time for good eating, enjoying time just to be. And when we recognise that the resources of our home, the Earth, are being over-exploited by excessive consumption and industrialisation, sooner or later a more frugal lifestyle will not only be desirable—it will become an imperative.

Also available from Green Books

Reduce, Reuse, Recycle!

An easy household guide
Nicky Scott, illustrated by Axel Scheffler
96pp £3.95 pb

This easy-to-use guide has the answers to all your recycling questions. Use its A-Z listing of everyday household items to see how you can recycle most of your unwanted things, do your bit for the planet, and maybe make a bit of money while you're at it.

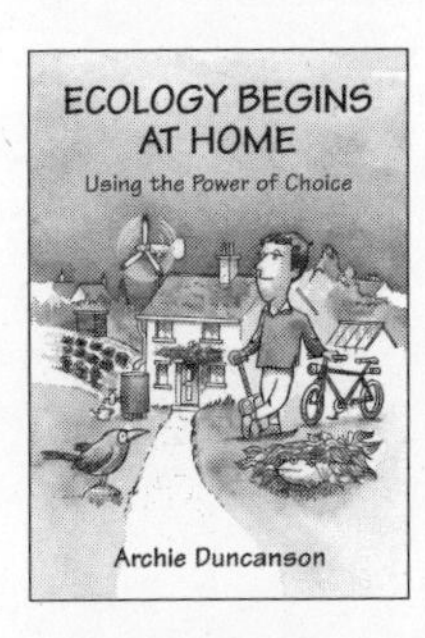

Ecology Begins at Home

Using the Power of Choice
Archie Duncanson
160pp, £4.95 pb

"Definitely the best book yet on how to green your lifestyle."—Permaculture Magazine

This book shows how one man looked around him and saw what he could do to reduce his personal ecological footprint. On the basis that you only need to take one step to make a difference, Archie takes you on his journey towards a more environmentally friendly home and an easier conscience. With delightful illustrations, and packed full of simple ideas to reduce the ecological impact of your daily life, *Ecology Begins at Home* is an inspiration for adults and children alike.

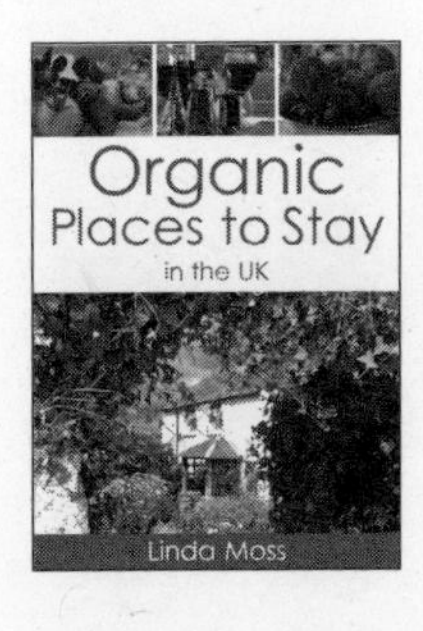

Organic Places to Stay in the UK

Over 500 places for an organic lifestyle away from home
Linda Moss
304pp in full colour £10.95 pb

Based on Linda Moss's successful website www.organicplacestostay.com, *Organic Places to Stay* is the first guide to B&Bs, small hotels and guesthouses in the UK that specialise in serving organic food. The guide is fully independent—no charge is made for entries. With over 500 entries, you can keep your organic lifestyle even when you are on holiday or visiting other parts of the UK.

. . . and many other books on green living, including eco-building and organic gardening. For a complete list, visit our website at www.greenbooks.co.uk